6th Asia Conference on Mechanical and Materials Engineering (ACMME 2018)

MATEC Web of Conferences Volume 213 (2018)

Seoul, South Korea
15 – 18 June 2018

Editors:

O.S. Es-Said
C.A. Huang

ISBN: 978-1-5108-7311-7

Printed from e-media with permission by:

Curran Associates, Inc.
57 Morehouse Lane
Red Hook, NY 12571

Some format issues inherent in the e-media version may also appear in this print version.

Printed by Curran Associates, Inc. (2018)

For additional information, please contact EDP Sciences – Web of Conferences
at the address below.

EDP Sciences – Web of Conferences
17, Avenue du Hoggar
Parc d'Activité de Courtabœuf
BP 112
F-91944 Les Ulis Cedex A
France

Phone: +33 (0) 1 69 18 75 75
Fax: +33 (0) 1 69 28 84 91

contact-edps@webofconferences.org

Additional copies of this publication are available from:

Curran Associates, Inc.
57 Morehouse Lane
Red Hook, NY 12571 USA
Phone: 845-758-0400
Fax: 845-758-2633
Email: curran@proceedings.com
Web: www.proceedings.com

TABLE OF CONTENTS

RESEARCH ON ASSEMBLY TECHNOLOGY OF ACTIVE MAGNETIC BEARING SYSTEM BASED ON GRAPH THEORY 1
Li Zhang, Yafeng Ren, Meiyun Yu, Jun Liu, Yanyan Luo

A STUDY FOR DIFFERENCES BETWEEN DIFFERENT DESIGN DRAWING PROGRAMS IN SHEET METAL DESIGN 5
Huang Yu-Che, Yeh Tien-Chaun

EFFECT OF GROOVE TEXTURE ON THE DYNAMIC CHARACTERISTICS OF WATER-LUBRICATED THRUST BEARING 9
Feng Huihui, Peng Liping

ANALYSIS AND DETECTION METHOD FOR MACHINING ERROR OF GLOBOIDAL INDEXING CAM PROFILE 14
Sun Shuwen, Jiang Zhengquan, Huang Jie, Yang Jianwu

ANALYSIS OF BRIDGE-TYPE DISTRIBUTED-COMPLIANCE MECHANISM 22
Lai Lei-Jie, Yin Xiao-Qia

ACCURACY SIMULATION METHOD BASED ON ASSEMBLY PROCESS INFORMATION MAPPING 26
Tang Jianjun, Liu Haitao

INVESTIGATION ON INFLUENCE OF CUTTING PARAMETERS ON SPINDLE VIBRATION OF CNC WOOD MILLING MACHINE 30
Van Thuy Tran, Nguyen Huu Loc

DETERMINING OPTIMAL GEAR RATIOS OF A TWO-STAGE HELICAL REDUCER FOR GETTING MINIMAL ACREAGE OF CROSS SECTION 36
Tuan Nguyen Khac, Pi Vu Ngoc, Thi Hong Cam Nguyen, Thao Tran Thi Phuong, Thanh Ho Ky, Hung Le Xuan, Tham Hoang Thi

CREATIVE DESIGN OF SITTING HUG MACHINE IN THE TREATMENT OF STUDENTS WITH AUTISM 41
Lo Jih-Sheng, Huang Shyh-Chour

INVESTIGATING PARAMETERS THAT EFFECT TO WEAR OF PLASMA NOZZLE 46
Pitayachaval Paphakorn, Sato Muhammatsoifu

DESIGN AND IMPLEMENTATION OF 3D PRINTING SYSTEM FOR CONTINUOUS CFRP COMPOSITES 50
Han Ningda, Cheng Jun, Yang Jiquan, Liu Yijian, Huang Wuyun

A REVIEW OF 3D FOOD PRINTING TECHNOLOGY 55
Pitayachaval Paphakorn, Sanklong Nattawut, Thongrak Anantapoom

A REVIEW OF PROCESS PARAMETERS THAT EFFECT TO EXTRUSION ON DEMAND 60
Pitayachaval Paphakorn, Baothong Thanakhan

A HIGH DEGREE OF CLEANLINESS OF THE ELECTRIC AIR SUPPLY VALVE DESIGN 64
Hu Luo

CHARACTERIZATION OF MORTAR FROM CHURCH RUINS IN BARANGAY BUDIAO, DARAGA, ALBAY 71
Mangay John Carlo A., Miranda Eric T., Anicas John Mark S., Recto Angel S., Cayme Jan-Michael C.

PETROGRAPHIC AND THERMAL CHARACTERIZATION OF MORTAR FROM A CHURCH RUIN IN CAGSAWA, ALBAY, PHILIPPINES 75
Laplana Nina Muriel A., Miranda Eric T., Mangay John Carlo A., Recto Angel S., Cayme Jan-Michael C.

STUDY ON THE ESTIMATION OF ROCK RHEOLOGICAL PARAMETERS UNDER MULTI-LEVEL LOADING AND UNLOADING CONDITIONS 79
Han Bing, Fu Qiang

SYNTHESIS AND CHARACTERIZATION OF POLYSULFONE (PSU)/PHILIPPINE HALLOYSITE (PH-HAL) NANOSTRUCTURED MEMBRANE VIA ELECTROSPINNING 84
Aquino Ruth R., Tolentino Marvin S., Arcamo Niel Karl G., Gara John Patrick N., Basilia Blessie A.

DENSITY OF TEST BODIES AND ITS EFFECT ON BURNING RATE OF FIRE-RETARDANT TREATED SAMPLES 89
Osvaldová Linda Makovická

SIMULATION AND EXPERIMENTAL STUDY ON FLOW AND HEAT TRANSFER OF SHIP CENTRAL COOLING SYSTEM 93
Fang Chao, Cai Biao-Hua, Ma Shi-Hua, Li Bin

A STATISTICAL APPROACH FOR FINDING INFLUENTIAL FACTORS IN RESPECT OF ENERGY CONSUMING OF A CAR PASSENGER 98
Ali Hussein M., Iqbal Asif
Author Index

Preface

It is our great pleasure to introduce the proceedings of 2018 6th Asia Conference on Mechanical and Materials Engineering (ACMME 2018) held in Seoul, South Korea from June 15-18, 2018. ACMME 2018 is dedicated to issues related to mechanical and materials engineering. One of the objectives of the conference is to establish platforms for collaborative research projects in this field, and to find potential opportunities for international cooperation.

The conference provides a platform to discuss Mechanical and Materials Engineering etc. with participants from all over the world, both from academia and from industry. Its success is reflected in the papers received, with participants coming from several countries, allowing a real multinational multicultural exchange of experiences and ideas.

The proceedings present a selection of papers submitted to the conference by universities, research institutes, and professionals working in various industries. All the papers were subject to peer-review by conference committee members and international reviewers. The papers were selected based on their quality and their relevance to the conference. The volume presents recent advances in the field of mechanical and materials engineering as well as various related areas, including materials science, biomaterials, and manufacturing processes, among others.

We would like to express our gratitude to all the members of the conference committee. We would also like to thank the reviewers, who spared their valuable time, for their advice. They have certainly helped improve the quality, accuracy, and relevance of each paper selected for the conference program and for publication. We also wish to thank all the authors who have contributed to this conference, as well as the organizing committee, reviewers, speakers, chairpersons, sponsors, and all the conference participants for their support for ACMME 2018.

Prof. Omar S. Es-Said, Loyola Marymount University, USA
29. 05. 2018

Statement of Peer review

In submitting conference proceedings to *Web of Conferences*, the editors of the proceedings certify to the Publisher that

1. They adhere to its **Policy on Publishing Integrity** in order to safeguard good scientific practice in publishing.
2. All articles have been subjected to peer review administered by the proceedings editors.
3. Reviews have been conducted by expert referees, who have been requested to provide unbiased and constructive comments aimed, whenever possible, at improving the work.
4. Proceedings editors have taken all reasonable steps to ensure the quality of the materials they publish and their decision to accept or reject a paper for publication has been based only on the merits of the work and the relevance to the journal.

Title, date and place of the conference

2018 6th Asia Conference on Mechanical and Materials Engineering
ACMME 2018
Seoul, South Korea
June 15-18, 2018

Proceedings editor(s):

Prof. Omar S. Es-Said, Loyola Marymount University, USA
Prof. Ching An Huang, Chang Gung University, Taiwan

Date and editor's signature

Research on Assembly Technology of Active Magnetic Bearing system Based on Graph Theory

ZHANG Li[1], REN Yafeng[2], YU Meiyun[1], LIU Jun[1], LUO Yanyan[1]
[1] *AVIC Qingan group co., LTD., 710077, Xian, China*
[2] *Northwestern Polytechnical University 710072, Xian, China*

Abstract. Active magnetic bearing system is a typical geometric constraint system which is combined with each part by certain assembly constraint relations. Graph theory is used to build the assembly model in the progress of assembly design. Trough the graph to express this system, it is not only clear and intuitive, and the algorithm is stable and high efficiency. Modeling active magnetic bearing assembly system is achieved by hierarchical model of data structure,which leads to accelerate the assessment of system configuration program and save a lot of time for designers.

1 Introduction

Active Magnetic Bearing (AMB) is a typical mechatronic system, which has no contact, wear and lubrication, and easy to controll its dynamic behavior and other series of characteristics[1]. The limitation of traditional mechanical bearing has been breakthrough, it has been widely attracted since it appears. In design of Active Magnetic Bearing system(AMBs), the component and configuration research is very important, and involves the spatial layout and assembly problem. Assembly design of AMBs is defined as dependent on the design method of predefined components or parts. Graph theory[2] are applied to build the assembly model of AMBs, it can quickly achieve the system configuration program. For the first time, Fudos and Hoffmann[3-5] have introduced the concept of cluster into the geometric constraint graph and solved the problem of assembly constraint by geometric constraint graph. N.abe[6] and others have developed a visualization system for assembly analysis and performance evaluation of mechanical system assembly. In the early stage of the design for assembly, this way that is advantageous to analyze the spatial structure and interference will led to accelerate the assessment of system configuration program, which save a lot of time for designers and improve the design efficiency.

2 Assembly Model of AMBs

Geometric constraint graph theory is the combination of graph theory and computer graphics. In the field of Graph theory, Geometric Constraint Graph（GCG）is a method that is used to model a geometric constraint system, which can directly manage or express the 2D or 3D space of entities and geometric constraints, and has good geometric space expanding and interpretability. In fact, AMBs is a typical geometric constraint system which is combined with each part by a certain assembly constraint relations. Through the graph to express system, it is not only clear and intuitive, and the algorithm is stable and high efficiency. Due to the magnetic bearing system with more parts and complex structures, the constraint relations of assembly model are complex. In terms of expression, maintenance and explanatory, the ideal assembly model has a consistency, which requires the establishment of effective assembly model.

The optimization process of assembly model of AMBs is set up in the process to turn relational model into a hierarchical model. Hierarchical model is consistency with the hierarchical structure of the product assembly in CAD system, which will be beneficial for modeling AMBs by CAD system. There is a kind of effective method to solve the assembly constraints of complex system. when AMBs is abstracted as geometric constraint graph , It is separated into solving the sub-problem, at the same time it is suitable for establishing strategy of solving the sub-problem. Eq. (1) is the mathematical expressions for geometric constraint system.

$$S = (X, C, A) \tag{1}$$

where, S is a Active Magnetic Bearing system. X is a collection of parametric parts. C is looked as a collection of geometric constraints. A is parameter values of geometric constraints. The following Eq. (2) is the mathematical expressions for geometric constraint graph.

$$GCG = (V, E) \tag{2}$$

GCG is a geometric constraint graph of AMBs. V is the GCG vertex and the same as X. Each vertex is a geometric entity. E is the line of GCG, same as C. Each line represents a geometric constraint sets. The assembly model is analyzed by five freedom AMBs, as shown in figure 1.The system is mainly composed of radial magnetic bearing(RMB), thrust magnetic bearing(TMB), sensor, motor, and other key parts[7].

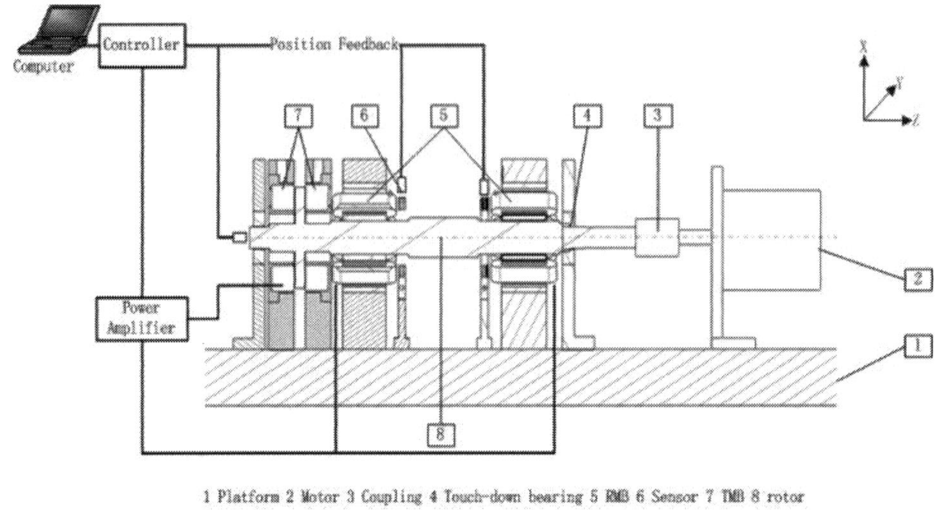

Figure 1. Five degrees of freedom AMBs

Figure 2. relational model and decomposing process diagram of AMBs.

Figure 2 is the relational model of five degrees of freedom AMBs. From figure 2a to figure 2c reflects the process of polymerization contraction between the assembly model. There are two kinds of nodes in the graph. Oval node represents component that contains modeling parameters, such as the stator or rotor , etc. Square node is the assembly constraint relations between components, for example C1 is an assembly constraint set between the central spindle and TMB rotor. The circle part that is dashed reflects a condensed and aggregated node called new node. It is the basic component of system structure. The component is an assembly unit to be achieved a certain function, which is a part or multiple parts.

In a new node, the distribution of constraints is changed and it results in the system assembly size to be reduced. It is called the decomposition process of geometric constraint system that separating from external constraints into the internal constraints. Specific speaking, C1, C12, C18, C24, C28 and C32 are converted into the rotor,s internal constraints. C3, C4 and C5 are converted into the internal constraints of TMB stator1. C8, C9 and C10 are converted into the internal constraints of TMB stator2. C14, C15 and C16 are converted into the internal constraints of RMB stator1.New node is replaced to form a new geometric constraint graph, it is a shown in the figure 2b. In the process of the second decomposition , the new graph can be integrated as a node, as shown in figure 2c.

From the decomposition process of relational model, hierarchical model of AMBs has be concluded. More importantly, the hierarchical model is used to model the assembly in CAD system, as shown in figure 3.This assembly model is divided into three layers. The first layer is the top level, which is as the general assembly. The second layer is seen as a sub- assembly. The last layer is the parts , which is the AMBs foundation.

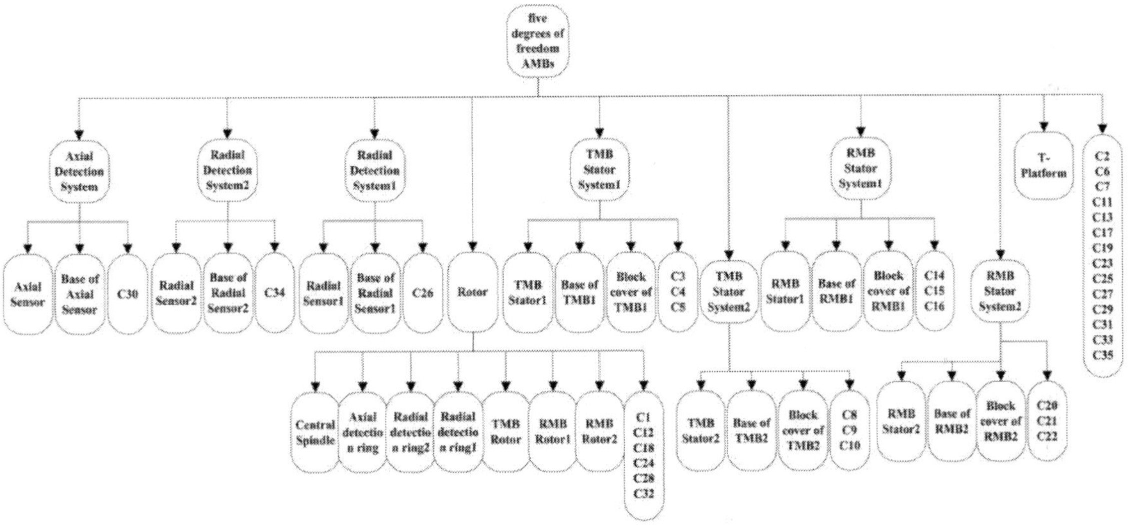

Figure 3. hierarchical model of AMBs.

3 Data Structure of AMBs

In the assembly model of AMBs, the node is relatively independent assembly unit, such as sub-assembly or component. The node is regarded as an object to define properties. The result is data structure of AMBs that includes assembly constraint relations, hierarchical structure and material properties. The following is a data structure.

a. Assembly:
   Name: the name of AMBs
   ID: the unique identification of AMBs
   Context: it contains all of Subassembly ID
   Pointer: it points to Subassembly

b. Subassembly:
   Name: the name of Subassembly
   ID: the unique identification of Subassembly
   Context: it contains all of Component ID
   Pointer: it points to Component

c. Component:
   Name: the name of Component
   ID: the unique identification of Component
   Context: it contains all of Part ID
   Pointer: it points to Part

d. Part:
   Name: the name of Part
   ID: the unique identification of Part
   Parameter: the parameters of Part
   Material: the material property of Part

e. Constraint:
   Name: the name of Constraint
   Type: the type of Constraint

Assembly is the top node that reflects the general assembly of AMBs, which includes Name, ID, Context and Pointer. The Pointer describes the direction of the next layer between assembly Constraints. The Subassembly is the second decomposition node. In fact, when the second aggregation is divided, designers interfere the subassembly in an interactive way, but to be following a certain principle. The component is a basic assembly unit of AMBs. In the interactive assembly, the component from a component library is used to mix design. The Part is the bottom of the basic component. The Constraint corresponds to the constraint in CAD system. it is advantageous to the depth of the secondary development.

4 Modeling of AMBs

Based on the assembly model and data structure , the input files for AMBs are designed .The corresponding windows are customized to realize the secondary development of CATIA by VB language. Program drives the input file to obtain the assembly information that are modeling parameters , constraints and material, etc. Basic component library and CATIA API interface are used to design a 3D model about AMBs. This 3D model is a preassembly model to be adjusted and modified. Figure 4 is the assembly modeling flowchart of AMBs.

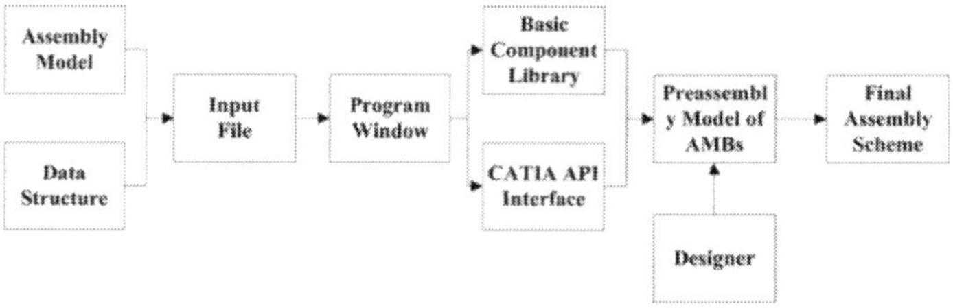

Figure 4. assembly modeling flowchart of AMBs.

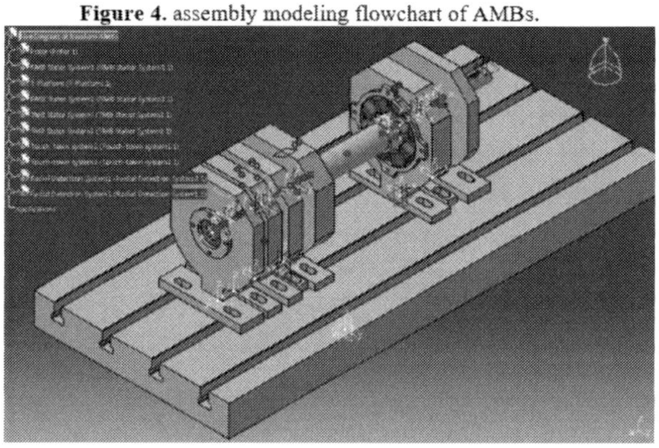

Figure 5. 3D model of AMBs.

According to the assembly modeling flowchart of AMBs, The top-down assembly design way is adopted to assembly 3D model in CATIA, as shown in figure 5.

5 Conclusion

Graph theory are applied to build the assembly model of AMBs, it can quickly achieve the system configuration program. In an early stage of the design for assembly, this way that is advantageous to analyze the spatial structure and interference will led to accelerate the assessment of system configuration program, which save a lot of time for designers and improve the design efficiency.

References

1. G.Schweitzer, H.Maslen. Magnetic bearing-Theory, design, and application to rational machinery [M], 2009
2. Koh Khee Meng. Introduction to Graph Theory [M], 2007
3. Fudos I, Hoffmann C M. A graph-constructive approach to solving systems of geometric constraints[J]. ACM Transaction on Graph, 1997, 16(2): 179~216
4. Bouma W, Fudos I, Hoffmann C M, et al. Geometric constraint solver[J]. Computer Aided Design,1995,27(6):487-501
5. Fudos I, Hoffmann C M. Correctness proof of a geometric constraint solver. Technical Report[J]. Dept of Computer Science , Purdue University, USA,1993
6. N.Abe,et al. Verification of assembly ability of mechanical parts and visualization of machinery of assembly in virtual space[J]. Proceedings of International Conference on Virtual Reality ,1997
7. Andrea Tonoli, Angelo Bonfitto, Mario Silvagni and Lester D. Suarez. Rotors on Active Magnetic Bearings[J]: Modeling and Control Techniques, 2012.

A study for differences between different design drawing programs in sheet metal design

Yu-Che Huang [1], Tien-Chaun Yeh [2]

[1] *Chaoyang University of Technology, Industrial Design Department, 168, Jifeng E. Rd., Wufeng Dist., Taichung 413, Taiwan (R.O.C.)*
[2] *Chaoyang University of Technology, Industrial Design Department, 168, Jifeng E. Rd., Wufeng Dist., Taichung 413, Taiwan (R.O.C.)*

Abstract. Most appearance and structure design has always been based on the product processing technology as the spindle, in general; sheet metal design changes in the appearance of the structure is based mainly on the basic structure of products and processing technology as the main core, and the appearance of manufacturing more Mechanical sheet metal, so most designers will use different design procedures to draw 3D graphics in the assembly or disassembly of parts, but different drawing procedures will affect the future assembly and disassembly of the design. In this study, we will compare two different models by building a model program using two different sheet metal structure drawings to present the problems that may be encountered in sheet metal design through a specific design flow. Through a real 3D drawing and design case, 3D graphics design flow, and finally through a design case to do a verification, to provide a reference for the future design.

1 Introduction

In the past, most mechanical sheet metal designs are based on the customer's design surface or 3D model as a design basis, and then through the R & D department or designer for part assembly or 3D parts rebuild, and then to meet the processing Factory processing, and designer identification. In addition, if the same customer needs to reduce costs, the product must have a certain similarity, must be little change, often only a small number of holes or the main dimensions need to be modified , If you want to re-disassemble the parts or 3D surface redesign, it is easy to waste time, and the process is also prone to design errors, just as "At the present time CAD is not a suitable tool for engineers. a long training and a fulltime commitment are necessary to be proficient with the parametric history-based modeling technology[1,2,3].

Engineers need an easy-to-use 3D tool tailored for conceptual design but usually they communicate concepts through 2D sketching or presentation tools and then a CAD operator has to interpret their wishes in a specific modeling tool. This way is easy to commit errors and lose time often resulting in not substantially achieved engineer effectively wanted ". So how to provide a good way of drawing is a topic worth discussing..

2 Literature

In the past, the establishment of a model based on the traditional history-based model took a great deal of time and mental effort. At present, a highly competitive enterprise can no longer exist. Therefore, Computer Aided Design (CAD) has been accompanied by computer operations since its introduction in the 1960s With the progress of capabilities and constant new developments, the model elements are modeled from the development of the wire architecture to the present, and from 2D to 3D in the space of the model. A new generation of modeling models (Direct modeling), also known as history-free model building, was first released by Space Claim Corporation in 2007, and then various vendors have also introduced the technology as The basic software[4,5,6,7], as shown in Table 2.

Among them, SIEMENS synchronous modeling technology introduced by Siemens in 2008 applied NX and Solid Edge software based on history-free direct modeling technology, providing designers with better design freedom. In addition, the software retains the traditional method of historical modeling, providing designers with the flexibility and choice to use. Combine it into two model building models. As shown in table 1:

Table 1. Different drawing methods.

Drawing way	(History-Based)	(History-free)
Establish a model approach (Design method)	(Order modeling)	(Synchronize modeling)

Table 2. Technical Development.

Technology	Product	Main features
Synchronous Technology	Siemens NX and SolidEdge	Hybrid technology: the direct modeler automate and control the history tree
Inventor Fusion Technology	Autodesk Inventor	Hybrid technology: is possible to switch between parametric and direct mode
CoCreate	PTC CoCreate	Direct modeling with the possibility to add constraint to the model
SolidWorks	Dassault Systemes SolidWorks	Hybrid technology: direct modeling features have been gradually added in SolidWorks
V6 direct editing	Dassault Systemes CATIA	Hybrid technology
SpaceClaim	SpaceClaim	Direct modeling
Iron CAD	Iron CAD	Hybrid technology

3 Research Methods

This study attempts to explore several issues: 1. History-Based modeling model and history-free model (History-Free) to establish a model approach in the design of sheet metal products are there any differences in the results of 2. History-based approach (History-Based) to establish model and history-free (History-Free) the establishment of model methods in the design of sheet metal products efficiency is no difference between the time 3. History-Based model of the establishment of the history and history (History- Free). The establishment of model approach in the design of sheet metal products feel no difference.

In this study, Solid Edge ST8 is used as a tool to design the design process, hoping to find a faster and more suitable design method. Help R & D or designers in the design of sheet metal applications. In addition, taking into account the operational limitations of the experiment, the choice of the electrical box as a case, are common sheet metal products, at the same time, in the experimental process to draw the model, the test will not cause too much difficulty.

4 Reserch Result

4.1 Start sheet metal design

In this part of the experiment, we want to explore the difference between synchronous modeling and order modeling when starting sheet metal design.

4.1.1 Start sheet metal design in synchronous mode

(Enter the part 1 to create the electrical box) Tools → Simultaneously build the model (Define the model used to build the model) (2) Center Point Rectangle → Smart Size → Stretch → Center Rectangle → Smart Size → Stretch Shell (To create the main body of the electrical box) (3) Center Point Rectangle → Smart Size → Mirror → Extrude → Center Rectangle → Smart Size → Mirror → Extrude → Linear → Smart Size → Mirror → Stretch (Excluding material) (Except for the model boundary to facilitate the follow-up model flat) (4) Center point and draw a circle → smart size → Stretch (Exclude) (after hole), 2D map of this study is shown in Figure 1 Show. Simultaneous establishment of model begin shown in Figure 1.

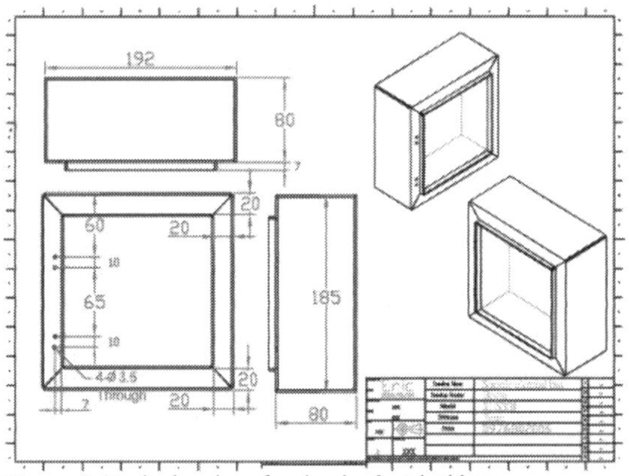

Figure 1. Begin drawings for simple electrical box.

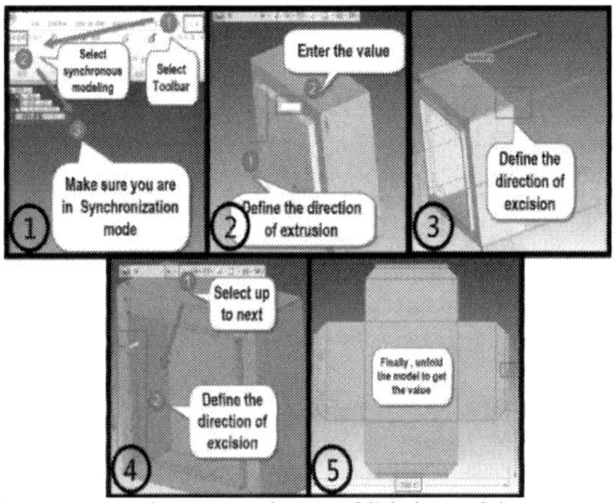

Figure 2. Synchronous mode to establish the model.

4.1.2 Start sheet metal design in order mode

Tools → Sequential Modeling (Define the model mode used) Sketch → Center Point Rectangle → Smart Size → Stretch → Sketch → Center Rectangle → Smart Size → Stretch → Thin Wall (Create Electrical Box Body) → Sketch → Center Point Rectangle → Smart Size → Mirror → Mirror → Stretch (Ext.) → Sketch Center Point Rectangle → Smart Size → Mirror → Mirror → Stretch (Ext.) → Sketch → Line → Smart Size → Mirror → Mirroring → Extrude (Excluding the model boundary for subsequent models) Basic drawing → Center point and

Circle → Extrude (Create hole position behind), as shown in Figure 3.

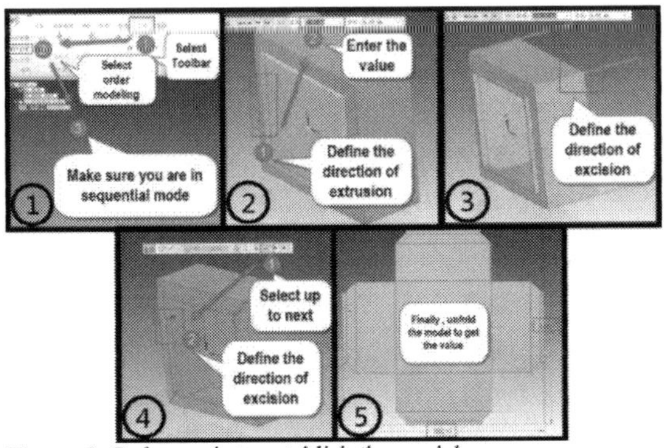

Figure 3. Order mode to establish the model.

4.1.3 The difference between the two

The simultaneous establishment of model and sequence model, the sheet metal design at the beginning of the construction, the instructions used in the difference is not too large, the most obvious difference is that the simultaneous model can be directly drawn in 3D space, and the order of the model Is the need to perform the drawing instructions, can sketch, in addition, the two finally expand the model, there is no difference in size, which means that both the model and the order of the establishment of the model to synchronize the two in the reconstruction of the sheet metal model , The final output of the model values and results no difference.

4.2 Sheet metal design in design change

In addition to the initial construction of the model in the sheet metal industry, design changes are also common issues in design. Therefore, it will be in the electrical box plate design changes in the site, and in different drawing patterns to compare the differences. The design changes in the site has the following 4 points:

(1) Electrical box length and width of the electrical box at the mouth(2)back hole diameter of the electrical box(3) Electrical box rear hole location(4)Electrical box height (plate thickness fixed).The following design changes when the sheet metal design reference map, as shown in Figure 4.

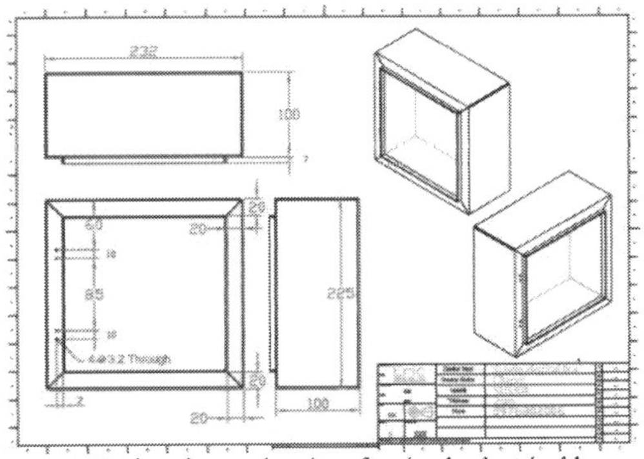
Figure 4. Design change drawings for simple electrical box.

4.2.1 Design changes in synchronous mode in sheet metal design

(1) select the function → modify the size → select the width of the frame part → modify the size.(2) Select four holes → modify the size. (3) Select the hole you want to move → Modify the size. (4) Select the upper part of the function keys → modify the size (this part of the instruction do not need to control).The relevant steps are shown in Figure 5.

Figure 5. Design chages in synchronous mode.

4.2.2 Design changes in order mode in sheet metal design

(1) into the sketch → modify the size (length and width changes)(2) into the sketch → modify the size.(3) into the sketch → modify the size.(4) into the stretch → modify the size (high modify) (this part requires instructions to do control).The related steps are shown in Figure 6.

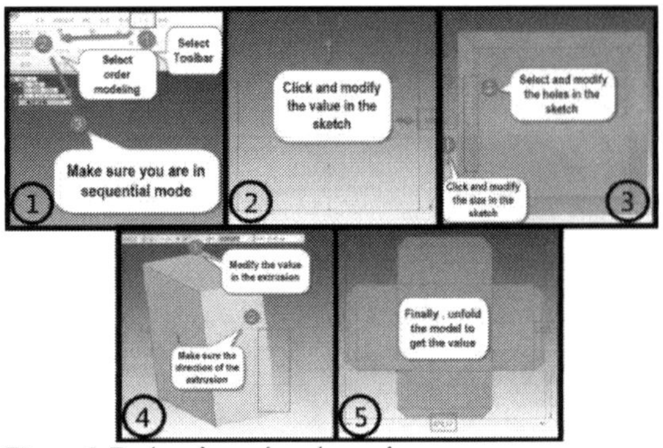

Figure 6. Design chages in order mode.

4.2.3 The difference between the two design changes.

Simultaneous modeling and sequential modeling, when the sheet metal design changes, synchronous modeling does not need to control through the instructions, and sequential modeling is based on the instruction back to the moon, to be modified, so if the order is built in the just When you start 3D rendering, there is no prior planning and future modifications can be a complex issue that can drastically reduce the designer's flexibility in designing. In addition, when both models are finally expanded, There is no difference in size, which means that both the synchronous model and the sequential model have no difference in the final design value of the sheet metal model. In designing, efficiency is the most important factor. In this study, the design process of Solid Edge ST8 was used as a planning tool. Subjects of industrial designers were tested and experimentally designed to give different subjects The modeling approach to the design process allowed the participants to draw the model based on pre-provided 2D maps and record the time it took. The test results are shown in the following Figure 7.

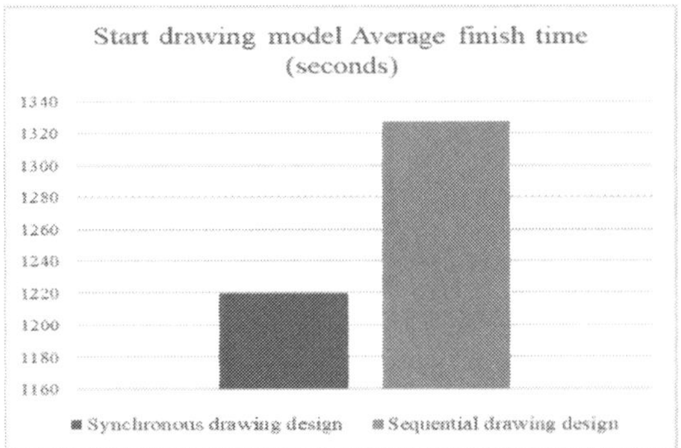

Figure 7. Start drawing model average finish time (seconds).

As you can see from the beginning of the design, the synchronous mode takes less time to complete, but the different result when the 3D design changes. The less time it takes for the sequential mode to be modeled, as shown in Figure 8.

In addition, a very time-consuming process of designing the work, the study used Solid Edge ST8 to plan the design flow. The experiment was conducted by the testers to draw 3D images and through two different design drawing methods Tests were designed to give participants different design approaches for modeling. Subjects were asked to draw 3D on the basis of the graphic content provided in this experiment. Then questionnaires were collected to collect the experience of the subjects using two different drawing and design methods. The test results are shown in Figure 8 below.

From the questionnaire of Figure 8 and related experimental results we can see that, whether it is in the beginning of the design or design changes, most of the subjects that synchronization modeling is easy to operate, the operation is also more efficient than sequential modeling, the other , The feeling of using it is also more relaxed.

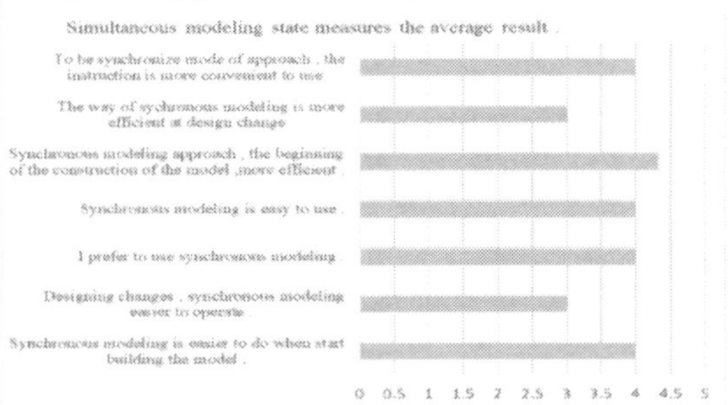

Figure 8. Simultaneous Modeling Attitude Scale Average.

References

1. Stefano Tornincasa , Francesco Di Monaco,The futureand the Evolution of CAD,Trends in the Development of Machinery and Associated Technology ,(2010)
2. Jenkins B. , 3D Engineering: How 3D Direct Modeling Empowers Conceptual Engineering and EnablesSimulation-Driven Product Development, White Paper,Ora Research LLC.(2009)
3. Chang Meng-Liang, A Study of Key Success Factors on Operational Performance for Sheet Metals Industry in Central Taiwan,(2014)
4. Gordon L., Comparing 3D CAD modelers, machinedesign.com,(2006)(http://machinedesign.com/article/comparing-3d-cad-modelers-1122)
5. Sunil V.B., Pande S.S., Automatic recognition of features from freeform surface CAD models, Computer-Aided Design **40**, 4 (2008), 502-517.
6. Sunil V.B., Agarwal R., Pande S.S., An approach to recognize interacting features from B-Rep CAD models of prismatic machined parts using a hybrid (graph and rule based) technique, Computers in Industry(in press), 2010.
7. Fiorentino M., Uva A.E., Dellisanti Fabiano M., Monno G., Improving bi-manual 3D input in CAD modeling by part rotation optimisation, Computer-Aided Design **42**,5(2010), 462-470.

MATEC Web of Conferences 213, 01003 (2018) https://doi.org/10.1051/matecconf/201821301003
ACMME 2018

Effect of groove texture on the dynamic characteristics of water-lubricated thrust bearing

Huihui Feng[1], Liping Peng[1]

[1]*Scholl of Mechanical and Electrical Engineering, Hohai University, Changzhou, Jiangsu 213022, China*

Abstract. In this study, the effects of groove texture on the dynamic characteristics of water-lubricated thrust bearing are theoretically investigated. The turbulent Reynolds equation and its perturbation equations for water-film lubrication are derived and solved by using finite difference method. Dynamic characteristics including the stiffness and damping coefficients of the bearing are calculated. The effects of rotary speed, film clearance and geometrical parameters including groove texture depth and circumferential angle on the dynamic characteristics of the bearing have been investigated.

1 Introduction

Water-lubricated bearings have attracted many attentions due to their low viscosity, low power loss at high rotary speed, good damping as well as long lift-time. Those bearings find application in water pumps, hydroelectric turbines and high-speed spindles [1]. However, despite of the low friction of the bearing at high rotary speed, the low viscosity also introduces a lower load capacity compared with the bearing lubricated by oil.

The application of surface texture has been shown to improve the performances of fully lubricated hydrodynamic bearing by many researchers [2,3]. The texture introduces micro-cavity effect which generates the additional film pressure of the bearing [2,3]. Tala-ighil et al. [4] pointed out that an appropriate arrangement of the textured area can improve the hydrodynamic load capacity and decrease the friction torque. Shi and Ni [5] found that with the increase of the groove number and reduction of groove size, the load capacity increased. As a result, in order to improve the load capacity of the water-lubricated bearing, the surface texture is introduced in this research.

In the previous study, the mathematical models and numerical solving methods were investigated to analyze the effect of rotary speed, shape of texture, texture depth and other structural parameters on the load capacity, friction of the bearings with surface texture. as the analysis of the bearings with surface texture were investigated [5-8]. Zhou et al. [6] analyzed the effects of the texturing parameters on the load capacity based on the Reynolds equation and proposed an optimal design method for the texture. However, the majority of the researches mainly focused on the static performances of the bearings with surface texture. So far, the dynamic characteristics of the bearings with texture have not been sufficiently investigated.

This paper aims to study the effect of groove texture on the stiffness and damping coefficients of hydrodynamic water-lubricated thrust bearing. The steady Reynolds equation considering turbulence is adopted; the linear perturbation method is used to derive its perturbation equations for water-film lubrication. And then the steady and perturbation equations are solved by finite difference method to obtain the steady and perturbed pressure distributions. Dynamic characteristics including the stiffness and damping coefficients of the bearing are calculated by integration of the pressure distribution. Finally, the effects of rotary speed and various geometrical parameters, such as groove texture depth, circumferential angle on the dynamic coefficients will be investigated.

Nomenclature	
c	Damping coefficient, N/(ms)
h_0	Film clearance, μm
h_{st0}	Maximum depth of a groove, μm
k	Stiffness, N/m
l_{st}	Width of a single groove
p	Dimensional pressure, Pa
G_θ, G_r	Turbulent coefficients
H	Dimensionless film thickness
H_{st}	Dimensionless depth of a groove
P	Dimensionless pressure
P_0	Reference pressure, Pa
μ	Viscosity, Pa·s
μ_0	Reference viscosity, Pa·s
θ	Angular coordinate
θ_{st}	Groove angle
Ω	Rotary speed, rpm

2 Mathematical Model

The schematic view of the groove-textured water-lubricated thrust bearing is shown in Figure. 1. The water was supplied through the inner side of the film and outflowed through the outer side of the film.

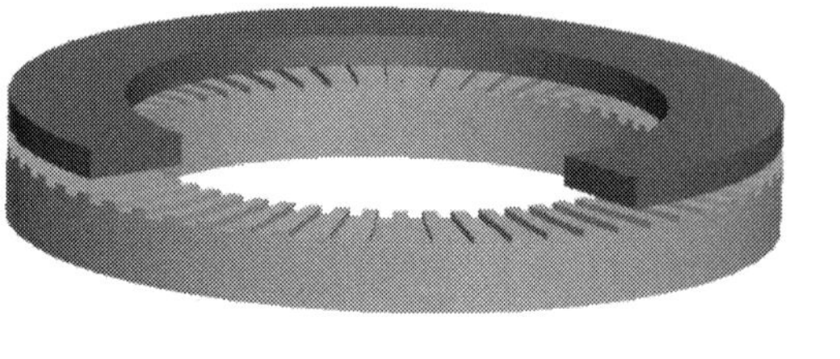

Figure 1. Schematic view of the groove-textured thrust bearing.

Both of the high rotary speed and low viscosity of water lead to large Reynolds number. In consequence, the water film becomes high turbulent. For an isoviscous incompressible fluid, the dimensionless Reynolds equation governing the turbulent flow for the water-lubricated thrust bearing is given as [9]:

$$\frac{\partial}{\partial R}\left(\frac{G_x R H^3}{\bar{\mu}}\frac{\partial P}{\partial R}\right)+\frac{\partial}{\partial \theta}\left(\frac{G_y H^3}{R\bar{\mu}}\frac{\partial P}{\partial \theta}\right)=\frac{R}{2}\frac{\partial H}{\partial \theta}+R\frac{\partial H}{\partial \tau} \quad (1)$$

Where $R=r/r_{in}$, $H=h/h_0$, $P=p/P_0$, $P_0=\Omega r_0^2\mu_0/c^2$, $\bar{\mu}=\mu/\mu 0$. The details of the turbulent coefficients G_r, G_θ can be obtained as follows [10]:

$$G_r=\frac{1}{12+0.0043\left(\mathrm{Re}_h\right)^{0.96}}$$
$$G_\theta=\frac{1}{12+0.0136\left(\mathrm{Re}_h\right)^{0.9}} \quad (2)$$

At both of the supply and exit plane of the bearing, the pressure takes a constant value equal to the ambient pressure.

$$P_{in}=P_{out}=1 \quad (3)$$

The Reynolds cavitation boundary condition is adopted.

The bottom line studied here is parabola, so the film thickness is expressed as follows:

$$H=\begin{cases}1 & \text{film land}\\ 1+H_{st} & \text{in the groove}\end{cases} \quad (4)$$

$$H_{st}=\begin{cases}0 & if\,(X,Y\notin\Psi)\\ -\frac{4h_{st0}}{l_{st}^2 h_0}X^2+\frac{h_{st0}}{h_0} & if\,(X,Y\in\Psi)\end{cases} \quad (5)$$

$$\Psi:-\frac{l_{st}}{2}\le X\le\frac{l_{st}}{2}\ and\ r_i\le Y\le r_o$$

Where, H_{st} is the dimensionless film thickness at the groove, r_i and r_o are the inner radial and outer radial of the thrust bearing, respectively.

Under the assumption of small amplitude motion of the thrust bearing, the perturbed form of the pressure expression and film thickness expression are substituted into the Reynolds equation to derive the zeroth and first-order of the disturbed Reynolds equation:

$$\frac{\partial}{\partial R}\left(\frac{G_r R H_0^3}{\bar{\mu}}\frac{\partial P_\xi}{\partial R}\right)+\frac{\partial}{\partial \theta}\left(\frac{G_\theta H_0^3}{R\bar{\mu}}\frac{\partial P_\xi}{\partial \theta}\right)=F_\xi,\ (\xi=0,z,\dot{z})$$
$$F_0=\frac{R}{2}\frac{\partial H_0}{\partial \theta}$$
$$F_z=-\frac{\partial}{\partial R}\left(\frac{3G_r R H_0^2}{\bar{\mu}}\frac{\partial P_0}{\partial R}\right)-\frac{\partial}{\partial \theta}\left(\frac{3G_\theta H_0^2}{R\bar{\mu}}\frac{\partial P_0}{\partial \theta}\right)$$
$$F_{\dot{z}}=R \quad (6)$$

The finite difference method is used to solve the Reynolds equation. The stiffness K_{zz} and damping coefficient C_{zz} of the thrust bearing can be calculated by integrating the perturbed pressure across the fluid film.

$$K_{zz}=-\iint \alpha P_z R d\theta dR \quad (7)$$
$$C_{zz}=-\iint \alpha P_{\dot{z}} R d\theta dR \quad (8)$$

3 Calculation procedure

The Reynolds equation is numerically solved by finite difference method (FDM). A computational MATLAB code is developed to calculate the performance of the bearing. The calculation process includes two main steps: static performance calculation and dynamic characteristics calculation. At first, the steady Reynolds equation is numerically solved to obtain the steady pressure distribution. The steady Reynolds equation refers to the equation (6) when the subscript of ξ satisfies ξ=0. Secondly, once the steady pressure distribution is obtained, the right part of the perturbed Reynolds equation (6) can be determined when the static pressure distribution P_0 is obtained. The FDM method is also used to discretize the perturbed equation.

4 Results and discussion

To investigate the effect of groove texture on the dynamic characteristics of the water-lubricated thrust bearing, the parameters of rotary speed, texture depth, and texture circumferential angle are selected. The parameters of the bearing and groove are listed in Table 1.

Table 1. Parameters of the groove-textured bearing.

Item	Value
Inner Diameter [mm]	35
Outter Diameter [mm]	50
Viscosity [Pa s]	0.001
Density [kg/m^3]	1000
Number of grooves	60

4.1. Effect of rotary speed

Figure 2 shows the predicted stiffness of the bearing with groove texture in respect to the rotary speed. The stiffness represents the ratio of force to the displacement. The depth and circumferential angle of a groove are 15μm and 3°, respectively. The film clearance h_0 is 15μm. As shown in the figure, the stiffness increases significantly with the rotary speed. This is expected, as the hydrodynamic effect induced by the groove texture improved with the increased rotary speed.

For damping coefficient shown in Figure 3, it increases gradually with the rotary speed.

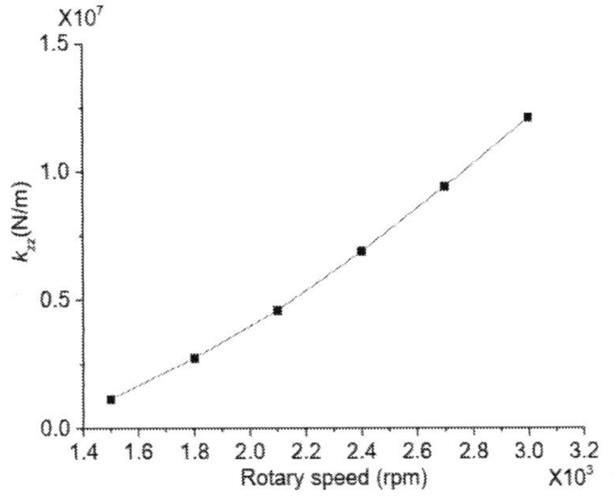

Figure 2. Stiffness vs rotary speed

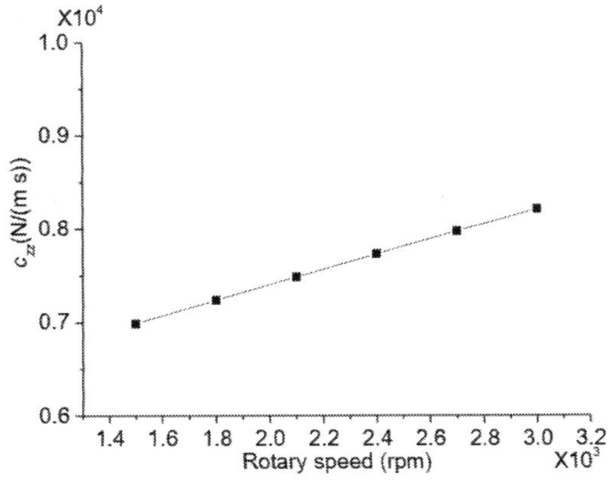

Figure 3. Damping coefficient vs rotary speed

4.2 Effect of film clearance

Figure 4 shows the effect of the film clearance on the stiffness of the bearing. The film clearance varies from 6μm to 21μm. The circumferential angle is 3 °.

According to the results, the stiffness decreases dramatically in respect to the film clearance. When the film clearance increases to 21μm, the stiffness is quite small. The reason for this is probably that the larger the film clearance is, the weaker the hydrodynamic effect is.

As shown in the Figure 5, the damping coefficient decreases from about 3×10^4N/(ms) to about 0.38×10^4N/(ms).

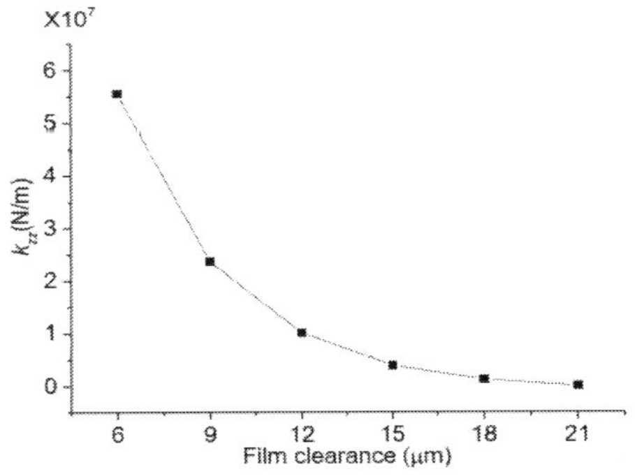

Figure 4. Stiffness vs film clearance

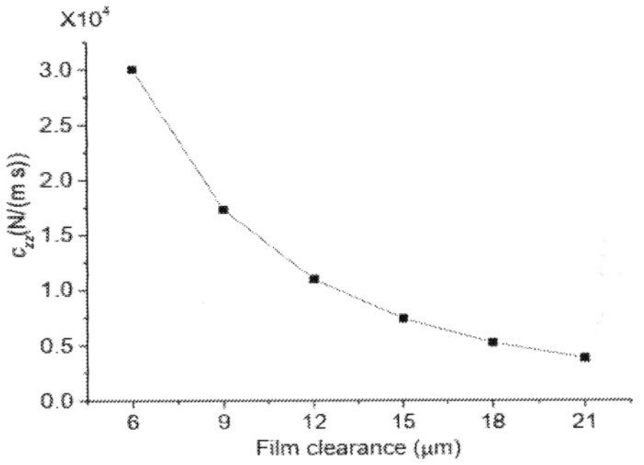

Figure 5. Damping coefficient vs film clearance

4.3 Effect of texture depth

Figures 6-7 illustrate the influence of the groove texture depth on the stiffness and damping coefficient of the bearing. The depth varies from 5μm to 25μm. The rotary speed is 20000rpm. The film clearance and circumferential angle of a groove are 15μm and 3°, respectively.

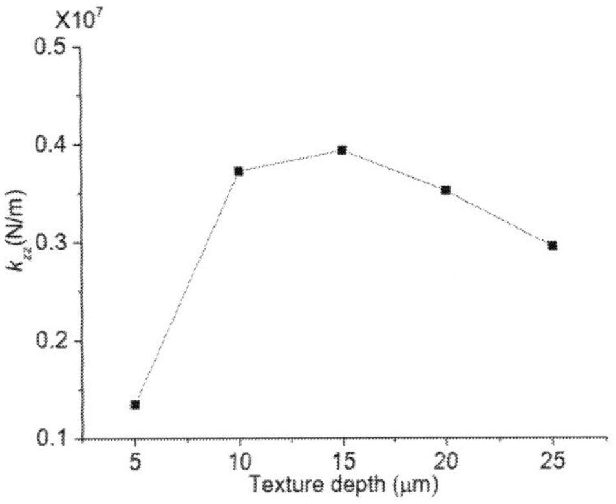

Figure 6. Stiffness vs texture depth

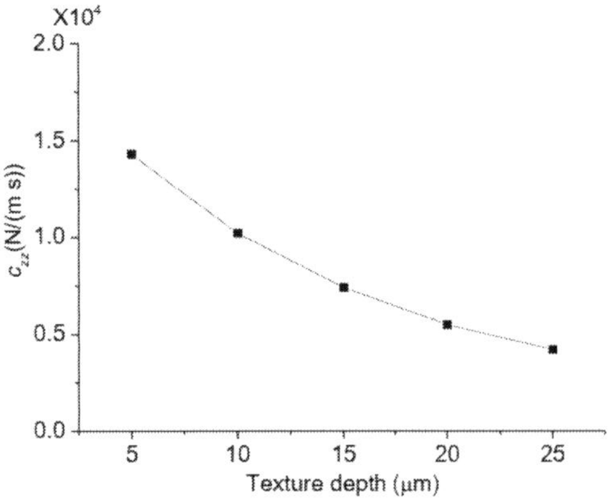

Figure 7. Damping coefficient vs texture depth

According to the results, the stiffness increases first and then decreases with the increasing texture depth. The reason for this is probably that the hydrodynamic effect increases first with the increasing texture depth; however, when the texture depth exceeds the film thickness, the drop of the pressure begins to be significant when the water flows from film land to texture. When the texture depth increases from 5μm to 10μm, there exists a significant increase for the stiffness; however, when the stiffness begins to decrease, the variation trend is slow. There exists a peak value when the texture depth is 15μm.

However, for the damping coefficient shown in Figure 7, it decreases gradually with the increased texture depth.

4.4 Effect of circumferential angle

The influences of circumferential angle on the dynamic characteristics of the bearing are illustrated in Figures. 8-9.

As shown in the figure, the circumferential angle is used to illustrate the texture width. Similar to the trend of the stiffness in respect to the texture depth, the stiffness increases first and then decreases with the circumferential angle. There is a peak value when the circumferential angle is equal to 4°.

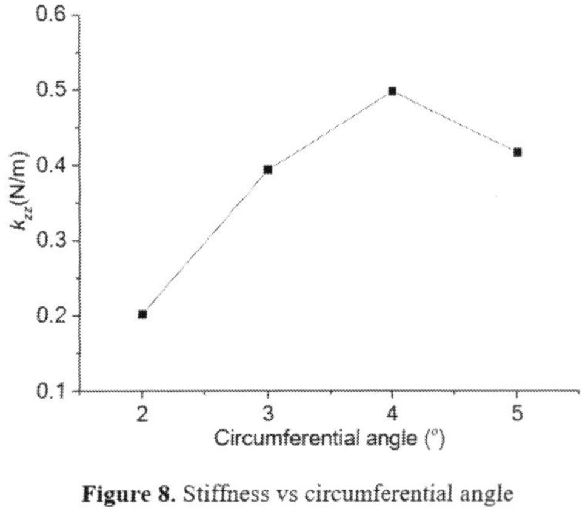

Figure 8. Stiffness vs circumferential angle

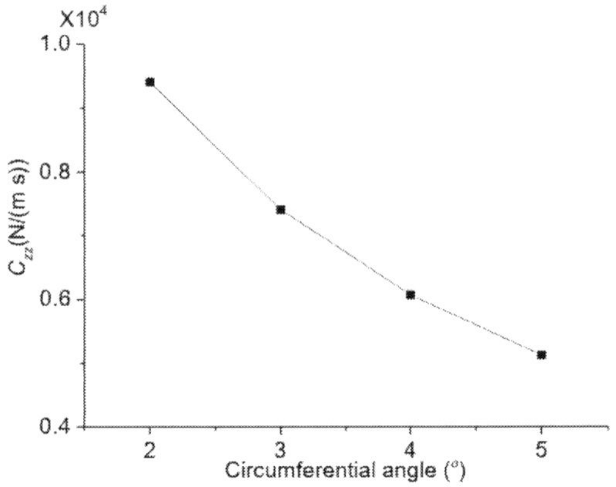

Figure 9. Damping coefficient vs circumferential angle

However, for the damping coefficient shown as in Figure 9, it still decreases in respect to the circumferential angle. With the circumferential angle increases from 2° to 5°, the damping coefficients decreases from 0.94×10^4N/(ms) to 0.52×10^4N/(ms).

5 Summary

The research carried out focused on the effect of groove texture on the stiffness and damping coefficients of water-lubricated thrust bearing. The turbulence is also included considering large Reynolds number lead by high rotary speed and low viscosity of water. The zeroth and first-order disturbed Reynolds equation to determine the dynamic characteristics and solved by FDM. The effects of rotary speed, film clearance, groove depth and circumferential angle on the dynamic characteristics of the bearing are analyzed. The main conclusions are summarized as follows:

(1) The stiffness and damping coefficients increase significantly with the increasing rotary speed due to the enhanced hydrodynamic effect.

(2) There exists an optimal texture depth and a circumferental angle for the stiffness of the bearing.

Acknowledgments

This work is supported by the National Natural Science Foundation of China (Grant NO. 51705131), the Fundamental Research Funds for the Central Universities (Grant NO.2016B14814), the Natural Science Foundation of Jiangsu Province (Grant NO. BK20160286) and the Changzhou Sci & Tech Program (Grant No. CJ20179022).

References

1. X.L. Zhang, Z.W. Yin, G.Y. Gao, Z. Li. Tribol. Int **85** 37 (2015)
2. D. Gropper, L. Wang, T.J. Harvey. Tribol. Int **94** 509 (2016)
3. T. Ibatan, M.S.Uddin, M.A.K.Chowdhury. Surf. Coat. Tech **272** 102 (2015)

4. N. Tala-Ighil, M. Fillon, P. Maspeyrot. Tribol. Int **44** 211 (2011)
5. X. Shi, T. Ni. Tribol. Int **44** 2022 (2011)
6. Y.K. Zhou, H. Zhu, W. Tang, C.B. Ma, W.Q. Zhang. Tribol. Int **52** 1 (2012)
7. N. Tala-Ighil, M. Fillon. Tribol. Int **90** 228 (2015)
8. J.Y. Zhang, Y.G. Meng. Tribol. Lett **46** 147 (2012)
9. H.H.Feng, S.Y.Jiang. P. I. Mech .Eng. C-J Mec **231** 3780 (2017)
10. S.M. Chun, D.H.Ha. Tribol. Int **34** 397 (2001)

Analysis and Detection Method for Machining Error of Globoidal Indexing Cam Profile

Sun Shuwen[1], Jiang Zhengquan[1], Huang Jie[1], Yang Jianwu[1]

[1]*No. 100, Pingleyuan, chaoyang district, Beijing, China*

Abstract. The profile of globoidal indexing cam is spatial undevelopable surface. It needs special CNC machine tool to finish batch production. The machining quality of the profile of globoidal indexing cam could be affected by the movement error of each part of machine tool and the clamped positioning error of workpiece. Firstly, this paper analyses the systematic error of the special machine tool of globoidal cam, derives the error transfer matrix, establishes the mathematical model of the globoidal cam profile which including all machining system errors through multi-body system theory primarily. Secondly,this method improves the measuring speed and avoids the probe radius compensation, and evaluates the tolerance of the profile greed of globoidal cam equidistant model directly. Finally,the feasibility of the measurement method is verified by experiments, and the experimental results are analysed and processed. It lays the foundation for the design, process and measurement of globoidal cam mechanism and provides a theoretical basis and key technology for developing the online rapid detection system of the globoidal indexing cam profile.

1 Introduction

The globoidal cam indexing mechanism has the advantages of compact structure, high indexing precision, strong bearing capacity, good high-speed performance and so on, so it has been widely used in ATC of CNC machine tools, tobacco machinery, packaging machinery. Globoidal indexing cam is a key component in Automatic Tool changer (ATC) [1-3], because ATC requires high precision of globoidal cam profile, the manufacture and tolerance detection of high-grade globoidal cam has been a difficult problem for domestic enterprises. Therefore, the profile accuracy of the globoidal indexing cam must be strictly guaranteed in the manufacturing process. It is of great practical significance to research the machining error anddetection of globoidal cam [4].

There are a number of scholars who have studied the machining error of globoidal cam. Wang [5] from National Chiao Tung University analyzes the influence on output accuracy of globoidal indexing cam mechanism caused by the installation error of three moving directions and three rotating directions in space. H. Y. Cheng [6] from Da-Ye University analyzes the error transfer of globoidal indexing cam mechanism systematically, establishing the error transfer equation, at the same time, he analyzes the sensitivity made by output precision to each error, and establishing the sensitivity equation. Yin Mingfu [7] from Tianjin University of Technology researches on the influence caused by center distance error of globoidal indexing cam machine tool on cam profile error and gives the changing trend of error. Ji Shuting [8] [9] from Beijing University of Technology researches on the influence caused by center distance error of the globoidal indexing cam special machining on the cam profile machining error and draws the influence coefficient curve, she reveals the result that the center distance error leads to the error of globoidal indexing cam profile. On the basis of the above research, the influence caused by two rotation shaft errors, three vertical errors and three linear displacement errors of globoidal indexing cam special machining on the machining error of globoidal indexing cam profile have been analyzed [10]. Wang Haiyan puts forward two kinds of scanning measuring method [11] for measuring the globoidal cam by applying of CMM. Tang Lin measures in the CMM and introduced the concept [12] of profile (surface) tolerance and helix tolerance. P.D.Lin and J.F.Hsieh researched on the conversion between the measurement coordinates and the theoretical position, to determine the measuring benchmark [13]. Led by the Jilin Industrial University, the precision index system [14] of globoidal cam mechanism is put forward. Song Lijuan puts forward the thought that reversely getting the manufacturing tolerance by measuring results, but she doesn't research on the specific contents [15]. Lin Xiaojun proposed the method to compensate [16] the radius by reconstructing the equidistant surface. However, the measuring speed of current CMM method is slow and complex, it cannot meet the measurement requirements of globoidal cam in production line, and it is difficult to fundamentally solve the problem of radius compensation.

In this paper, based on the working principle of the globoidal cam machine, the mathematical model of the error of globoidal cam profile is established by using the multi-body system theory. A fast measuring method based on the equidistant model characteristic line of globoidal cam is proposed. This method can be used in the measurement of production line, and the measuring speed is fast, so the radius compensation is avoided effectively.

2 Analysis of Cam Special Machine toolError

2.1 Processing Principle of Globoidal Cam

The whole structure of special CNC machine tool for globoidal indexing cam is shown in Fig.1a, it uses four axes linkage pattern and generating method to process. The four axes are represented by "*A*", "*B*", "*W*" and "*Z*" respectively. Axis *A* and axis *B* are rotating axes, axis*W* and axis*Z* are linear moving axes. While machining globoidal indexing cam, it should not only rotates around its own axis but also swing. The two motions are realized and controlled by the motion of the mutually interactive axis *A* and *B* respectively; the distance between axis *A* and axis *B* is the center distance of globoidal cam indexing mechanism which is adjusted and controlled by axis*W*; axis *Z*realizes the linear feed motion of control cutter along its central axis. The relative motion between workpiece and cutter realized by the linkage of axis *A*and axis*B* is equivalent to the pre-designed relative motion between globoidal indexing cam and driven turntable in the globoidal cam mechanism.

2.2 Definition of Cam Special Machine Tool Error

It is necessary to consider the many kinds of machine tool errors in the manufacturing process of the globoidal indexing cam. According to the multi-body theory, the motion error model is shown in Fig.1. At this time, there are two branches in the machine tool processing system.

Cutter branch: machine tool body----spindle sliding table axis*Z*----spindle----cutter;

Cam branch: machine tool body----turntable axis*B*----powered support axis*W*----rotary axis*A*----globoidal cam.

There are three linear displacement errors and three angular displacement errors in the spindle, axis *A*, axis *B*, axis *Z* and axis *W*. And it is also necessary to consider the error of cam rotation angle and driven turntable rotation angle and the centre distance error between camshaft and driven shaft, since there is relative motion relation between the cam rotation angle and the driven turntable rotation angle. In order to facilitate understanding, describing the above error decomposition diagram according to the machine tool structure in Fig.1, as shown in Fig.2. The error definition of globoidal cam machine too and symbolic representation are shown in table 1.

Fig.2 defines the error coordinates of moving parts of the globoidal indexing cam special machine tool. Fig.2ais themachine tool bed, Fig.2b is axis *B*, Fig.2c is axis *W*, Fig.2d is axis *A*, Fig.2e isthe spindle,Fig.2f is axis *Z*.

2.3 Feature Transformation Matrix.

The feature transformation matrix of multi-body system is represented by homogeneous coordinate. Homogeneous coordinate transformation includes six transformation modes: linear displacement along three coordinate axes and angular displacement around three coordinate axes. According to the theory of homogeneous coordinate transformation, we can get each feature transformation matrix between each moving body and its adjacent low-sequence body. Therefore, the transformation matrix of error coordinate system of spindle error of machine tool, the axis B, rotating axis A, power support axis W and the spindle sliding table axis Z can be obtained. The transformation matrix includes linear displacement matrix and angular Displacement matrix.

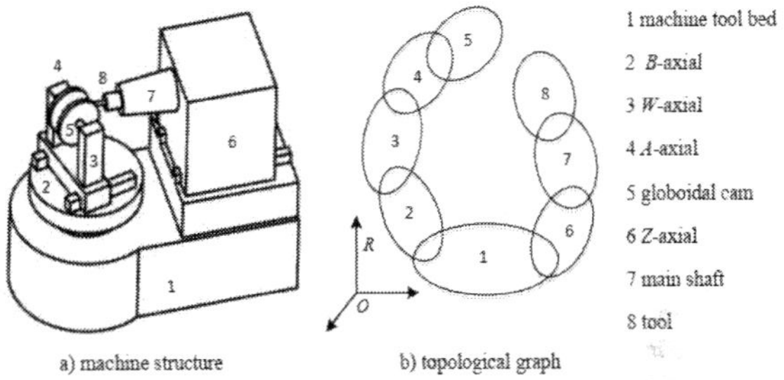

Figure1. Topology diagram of special machine tool

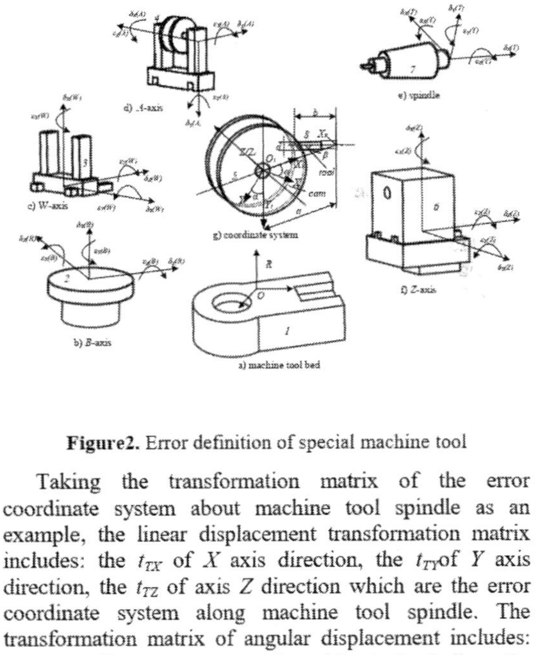

Figure2. Error definition of special machine tool

Taking the transformation matrix of the error coordinate system about machine tool spindle as an example, the linear displacement transformation matrix includes: the t_{TX} of *X* axis direction, the $t_{TY}$of *Y* axis direction, the t_{TZ} of axis *Z* direction which are the error coordinate system along machine tool spindle. The transformation matrix of angular displacement includes: error coordinate system around machine tool spindle----the r_{RX} which rotates around *X* axis, the r_{RY} which rotates around the *Y* axis, the r_{RZ} which rotates the axis *Z*.

$$t_{TX} = \begin{bmatrix} 1 & 0 & 0 & \delta_X(T) \\ 0 & 1 & 0 & 0 \\ 0 & 0 & 1 & 0 \\ 0 & 0 & 0 & 1 \end{bmatrix} \qquad (1)$$

$$t_{TY}=\begin{bmatrix}1&0&0&0\\0&1&0&\delta_Y(T)\\0&0&1&0\\0&0&0&1\end{bmatrix} \quad (2)$$

$$t_{TZ}=\begin{bmatrix}1&0&0&0\\0&1&0&0\\0&0&1&\delta_Z(T)\\0&0&0&1\end{bmatrix} \quad (3)$$

$$r_{RX}=\begin{bmatrix}1&0&0&0\\0&\cos(\varepsilon_X(T))&\sin(\varepsilon_X(T))&0\\0&-\sin(\varepsilon_X(T))&\cos(\varepsilon_X(T))&0\\0&0&0&1\end{bmatrix} \quad (4)$$

$$r_{RY}=\begin{bmatrix}\cos(\varepsilon_Y(T))&0&-\sin(\varepsilon_Y(T))&0\\0&1&0&0\\\sin(\varepsilon_Y(T))&0&\cos(\varepsilon_Y(T))&0\\0&0&0&1\end{bmatrix} \quad (5)$$

$$r_{RZ}=\begin{bmatrix}\cos(\varepsilon_Z(T))&\sin(\varepsilon_Z(T))&0&0\\-\sin(\varepsilon_Z(T))&\cos(\varepsilon_Z(T))&0&0\\0&0&1&0\\0&0&0&1\end{bmatrix} \quad (6)$$

Similarly, the linear displacement and angular displacement matrix of error coordinate system of turntable axis *B*, the rotating axis *A*, the power support axis *W* and spindle slide table of *Z* can be obtained respectively. The representation method of the transformation matrix of each axis about the machine tool are shown in table 2.

Table1. Errors of the special machine tool

Error Type	Linear offset error			Rotational error		
Error Direction	*X*	*Y*	*Z*	*X*	*Y*	*Z*
Spindle	$\delta_X(T)$	$\delta_Y(T)$	$\delta_Z(T)$	$\varepsilon_X(T)$	$\varepsilon_Y(T)$	$\varepsilon_Z(T)$
Z-axis	$\delta_X(Z)$	$\delta_Y(Z)$	$\delta_Z(Z)$	$\varepsilon_X(Z)$	$\varepsilon_Y(Z)$	$\varepsilon_Z(Z)$
B-axis	$\delta_X(B)$	$\delta_Y(B)$	$\delta_Z(B)$	$\varepsilon_X(B)$	$\varepsilon_Y(B)$	$\varepsilon_Z(B)$
W-axis	$\delta_X(W)$	$\delta_Y(W)$	$\delta_Z(W)$	$\varepsilon_X(W)$	$\varepsilon_Y(W)$	$\varepsilon_Z(W)$
A-axis	$\delta_X(A)$	$\delta_Y(A)$	$\delta_Z(A)$	$\varepsilon_X(A)$	$\varepsilon_Y(A)$	$\varepsilon_Z(A)$
Cam Angle	--	--	--	--	--	$\Delta\alpha$
Turntable Angle	--	--	--	--	--	$\Delta\beta$
Center Distance	Δa	--	--	--	--	--
Turntable	$\delta_X(S)$	$\delta_Y(S)$	$\delta_Z(S)$	$\varepsilon_X(S)$	$\varepsilon_Y(S)$	$\varepsilon_Z(S)$

Table2. Representation method of transformationmatrix

Coordinate System	Linear Displacement			Angular isplacement		
	X	*Y*	*Z*	*X*	*Y*	*Z*
Spindle	t_{TX}	t_{TY}	t_{TZ}	r_{RX}	r_{RY}	r_{RZ}
B-axis	t_{BX}	t_{BY}	t_{BZ}	r_{BX}	r_{BY}	r_{BZ}
A-axis	t_{AX}	t_{AY}	t_{AZ}	r_{AX}	r_{AY}	r_{AZ}
W-axis	t_{WX}	t_{WY}	t_{WZ}	r_{WX}	r_{WY}	r_{WZ}
Z- axis	t_{ZX}	t_{ZY}	t_{ZZ}	r_{ZX}	r_{ZY}	r_{ZZ}
Cam Angle	--	--	--	--	--	r_α
Turntable Angle	--	--	--	--	--	r_β
Center Distance	--	--	t_a	--	--	--

When deriving the globoidal indexing cam profile equation, there are three rectangular coordinate systems are established, as shown in Fig.2. The coordinate system $O_0X_0Y_0Z_0$ is a fixed coordinate system, its origin O_0 is the intersection point between the center axis of cutter and the swinging centerline of the workpiece, that is, the intersection point of the center axis of the axis*Z* and the axis*B*. The O_0Z_0 axis overlaps with the centreline of cutter, that is, overlapping withaxis*Z*. The O_0Y_0 axis overlaps with the swinging centerline of the workpiece, that is, overlapping with the centerline when axis*B*move. The coordinate system $O_1X_1Y_1Z_1$oscillates around the axis*B* and theO_1Z_1 axis overlaps with the workpiece rotation centreline, that is, it overlaps with the centerline whenaxis *A*move; The O_1X_1 axis overlaps with the line O_1O_0, that is, parallel with the axis*W*. And the coordinate system is fixed with workpiece. Making the coordinate system$O_1X_1Y_1Z_1$ rotates around the O_1Z_1 axis, the rotation angle is equal to the angular Displacement of axis*A* movement, then we can get the coordinate system *OXYZ*. The point *P*on the centerline of cutter is a random point on the theoretical profile after processing. The theoretical profile of globoidal indexing cam is generated by using point*P*and through the generating method. In Fig.2, *b* is the distance between the point *P* and the point O_0, *d* isthe diameter of cutter, *a* is the distance between theaxis*A* rotation center line and the axis*B* swing centerline of machine tool, that is the distance between cam rotation center axis line and the cam swing center axis line, which is called center distance. The relation between αandβ is determined by the motion law of a pre-designed globoidal indexing cam mechanism, which is a known condition.

According to the coordinate system established in Fig.2 and the feature transformation matrix based on multi-body theory, the position vector $p_0$of point *P* in coordinate system $O_0X_0Y_0Z_0$is as follows:

$$p_0=\begin{pmatrix}x_0\\y_0\\z_0\end{pmatrix}=\left(t_{BX}t_{BY}t_{BZ}r_{BX}r_{BY}r_{BZ}\right)\left(t_{ZX}t_{ZY}t_{ZZ}r_{ZX}r_{ZY}r_{ZZ}\right)\left(t_{TX}t_{TY}t_{TZ}r_{TX}r_{TY}r_{TZ}\right)t_b \quad (7)$$

The position vector p_1 of point *P* in coordinate system$O_1X_1Y_1Z_1$is as follows:

$$p_1=\begin{pmatrix}x_1\\y_1\\z_1\end{pmatrix}=\left(t_{WX}t_{WY}t_{WZ}r_{WX}r_{WY}r_{WZ}\right)r_\beta p_0+t_a \quad (8)$$

The position vector *p* of point *P* in the coordinate system *OXYZ* is as follows:

$$p=\begin{pmatrix}x\\y\\z\end{pmatrix}=t_\alpha p_1 \quad (9)$$

The unit normal vector of theoretical profile after machining is as follows:

$$n = \frac{\frac{\partial p}{\partial b} \times \frac{\partial p}{\partial \alpha}}{\left| \frac{\partial p}{\partial b} \times \frac{\partial p}{\partial \alpha} \right|} \quad (10)$$

The actual profiles on both sides of the processing are represented by p_l and p_r respectively, and the solution is as follows:

$$p_l = p - \frac{d}{2} n \quad (11)$$

$$p_r = p + \frac{d}{2} n \quad (12)$$

Adding the equation (9), (10) to the equation (11), (12), the error model of globoidal indexing cam profile could be obtained.

Setting all the error value in table 1 as 0, then the theoretical profile equation of globoidal indexing cam could be obtained.

3 Measuring Methods of Cam Profile

3.1 MeasuringMethod of Equidistant Surface.

Only when the ruby measuring ball of CMM contacts with the object to be tested, the position information (coordinate) of the ruby-measuring ball will be recorded. Therefore, the distance between the coordinate and the contact point is the radius R of ruby measuring ball; All the points on the object to be tested are recorded by CMM, and all the distance from these points to the object under test are the radius R of ruby measuring ball, when we fit these points into a complete surface, the equidistant model of the object to be tested is obtained, and the distance between the equidistant model and the object to be tested is R. As shown in Fig.3.

3.2 Modeling Method of Equidistant Surface.

According to the principle of Equidistant surface, making the point on the theoretical profile deviate along its normal direction, and keeping the deviating distance is equal to the radius of cutter, so that the practical profile can be obtained.

The globoidal cam mechanism is composed of globoidal cam and follower (roller). The motion law of globoidal cam and follower is modified sine motion law, according to the profile equation given in design and application innovation8 of intermittent motion mechanism, the equidistant model of the contact point between cam and follower extending along its normal vector direction is equivalent to the globoidal cam profile equation by reducing the radius of the follower, as shown in Fig.4.

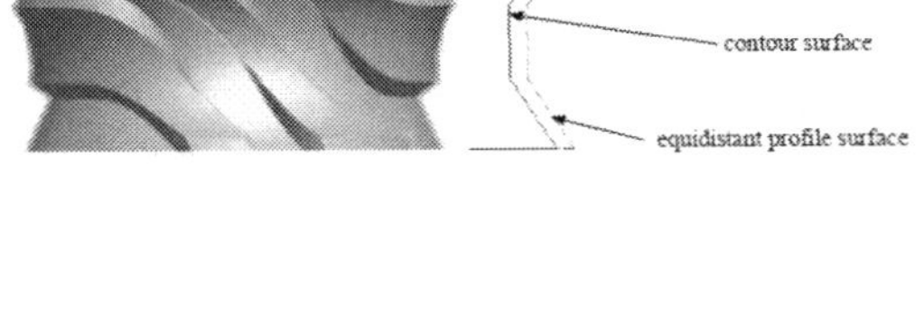

Figure3. Equidistant profile model theory

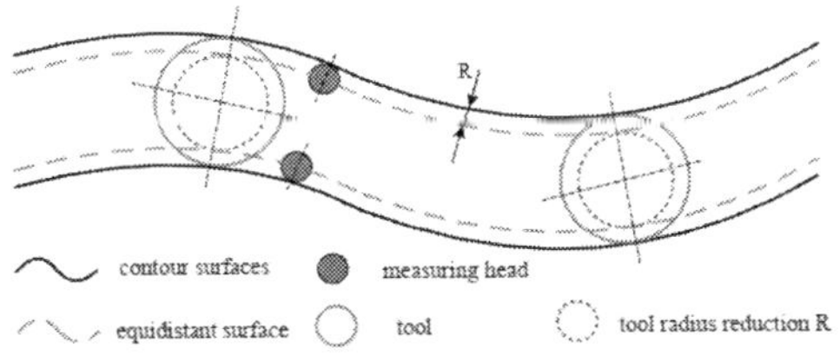

Figure4. Relationship of equidistant model and cutter radius

Taking the globoidal cam as the starting pointand then modifying its theoretical model so as to construct the equidistant model. The measuring result of the CMM is the spherical center coordinate of the measuring ball, and the distance between the contact point and the spherical center is the radius R of the measuring ball. According to the principle of equidistant surface, making every point P of the globoidal cam deviate along its normal vector direction n, and deviating the fixed distance R, then we can get the corresponding equidistant points of P, that isP_l' andP_r', which are used as the theoretical equidistant profile model for the coordinate measuring.

$$\boldsymbol{P}_l' = \boldsymbol{P} - \left(\frac{d}{2} - R \right) n \quad (13)$$

$$\boldsymbol{P}_r' = \boldsymbol{P} + \left(\frac{d}{2} - R \right) n \quad (14)$$

3.3 Analysis of Measuring Principle.

In order to improve the measuring speed and evaluate the line profile tolerance, the method of measuring the characteristic line of the equidistant surface model is put forward, as shown in Fig.5.

No matter where the contact point of CMM measuring ball and globoidal cam is, only keeping the spherical center of measuring ball moves in the specific axial section, so the trajectory of spherical center must in this axial section, the ideal trajectory of the spherical center is the intersection line of the theoretical equidistant surface and the axial section, thus we can use the characteristic line of equidistant model axial section as the evidence of evaluating the line profile of globoidal cam, and there is no need to compensate the probe radius. In the actual measurement, it is only necessary to make the probe of CMM move along the direction of axial section and just making sure the movement of the spherical center in corresponding axis section of this characteristic line. This measuring method could improve the measuring speed through unchanging the angle of the probe.

The axial section equation atα angle is as follows:

$$y / x = \tan(\alpha) \quad (15)$$

According to the definition of the equidistant profile of globoidal cam and the characteristic line of the axial section, we can form an equation sets through combining the formula(13) or formula(14) of equidistant profile of globoidal cam respectively with axial equation(15) to form a simultaneous equation, and the result is the characteristic line of equidistant profile model.

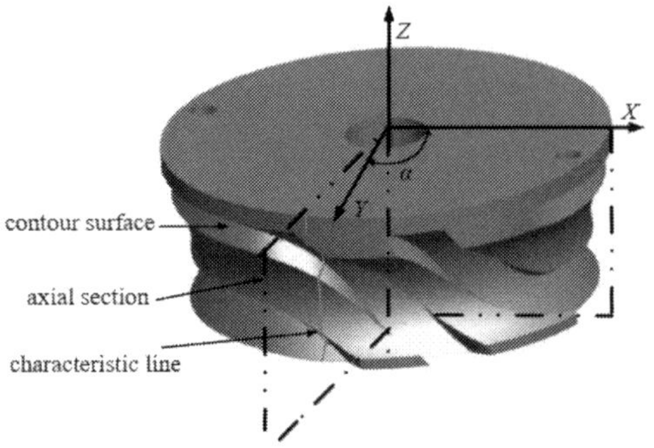

Figure5. Characteristic line diagram

3.4 Construction of Measurement System.

Using the high-precisionCMM to build precision turntable so as to form the four axis measurement system, achieving rapid measurement of tolerance about globoidal cam profile, the overall structure of the measurement system is shown in Fig.6. The measurement system is composed of CMM, precision turntable, motor controller, driver and digital display. The digital display shows angle information of turntable in real time, processing the information and the measuring results through computer so as to improve the measuring speed and accuracy.

4 Measurement Experiment and Analysis

4.1 Composition of Measurement System.

The whole set of device of tolerance detection system about globoidal indexing cam profile as shown in Fig.7, the specific experimental conditions includes: CMM, model NHC-Y564, measurement range is X axis -500mm , Y axis -600mm, Z axis -400mm; precision turntable system includes: the turntable and the controller, circular grating angle measurement system and the positioning module; the work piece be measured, TC40 type of globoidal indexing cam, size: thickness of 97mm , maximum diameter of 265mm.

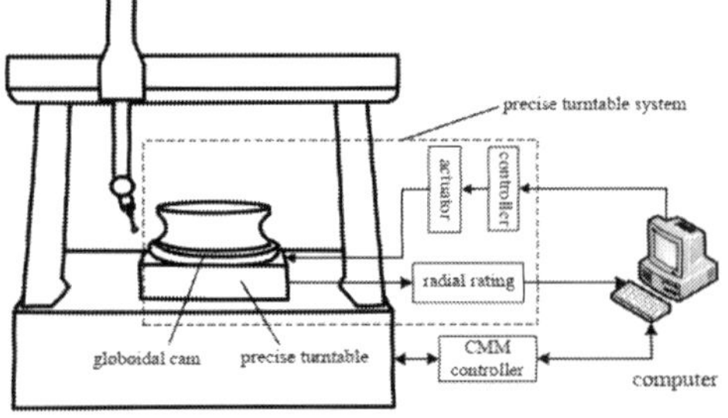

Figure6. Structural diagram of measuring device

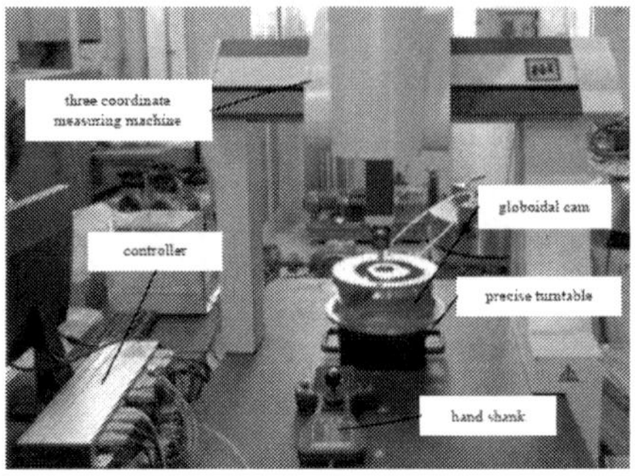

Figure7. Set-up diagram of measurement system

4.2 Process of Measurement Experiment.

Taking the globoidal cam profile equidistant model as the theoretical model, based on the manufacturing datum of the globoidal cam, establishing the datum coordinate system, which makes the globoidal cam and its theoretical model in the same coordinate system. During the measurement process, after a characteristic line has been measured, controlling the precision turntable rotates to the next characteristic line which is going to be measured. Combining the measuring method of the globoidal cam equidistant model with the above referred device, the measuring process is as follows.

Firstly, establish a reference coordinate system. Secondly, found a measurement coordinate system. Thirdly, measuring the feature points while adjusting the turntable angle. Export data of feature points at the end of measurement.

Byselecting different measuring angles, the equidistant characteristic point of the corresponding angle can be measured, and comparing it with the theoretical characteristic line under the same angle, calculating to get the line profile tolerance of the globoidal cam equidistant model .

4.3 Analysis of Measurement Results.

Taking this paper takes the TC40 globoidal cam as an example, the data point coordinates obtained from the CMM are shown in Table3:

The data in table 1 are the characteristic point coordinates of the globoidal cam intermittent section and PT1-PT8 is the characteristic line of the upper part in Fig.5; PT9-PT16 is the characteristic line of the lower part of Fig.5.

Calculating the distance of actual measuring characteristic point Pt by using Matlab according to the method described in 3, the results are shown in Fig.8.

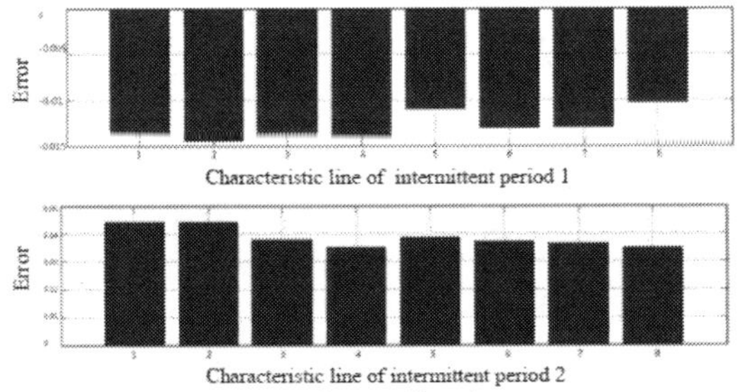

Figure8. Line profile tolerance

The maximum value Dtmax from the characteristic point Pt to the theoretical characteristic line Si is 0.043 mm and the minimum value of Dtmin is -0.014 mm.

$$\Delta L = D_{tmax} - D_{tmin} = 0.043 - (-0.01) = 0.057\text{mm} \quad (10)$$

Therefore, the line profile tolerance of the characteristic line under the cam angle is ΔL=0.057mm.

Table3. Coordinates of characteristic points on dwell segment

Point	Coordinate	Theory (mm)	Practical (mm)
PT1	X	122.2390	122.2735
	Z	-18.5961	-18.6978
PT2	X	122.9901	123.0258
	Z	-19.8971	-20.0028
PT3	X	123.7412	123.7742
	Z	-21.1980	-21.2973
PT4	X	124.4923	124.5281
	Z	-22.4989	-22.6038
PT5	X	125.2433	125.2835
	Z	-23.7999	-23.9063
PT6	X	125.9944	126.0328
	Z	-25.1008	-25.2083
PT7	X	126.7455	126.7817
	Z	-26.4017	-26.5053
PT8	X	127.4966	127.5366
	Z	-27.7026	-27.8074
PT9	X	107.5998	107.6338
	Z	-73.2422	-73.3760
PT10	X	108.9003	108.9354
	Z	-72.4913	-72.6245
PT11	X	110.2007	110.2345
	Z	-71.7405	-71.8670
PT12	X	111.5012	111.5381
	Z	-70.9897	-71.1115
PT13	X	112.8017	112.8387
	Z	-70.2389	-70.3646
PT14	X	114.1021	114.1348
	Z	-69.4880	-69.6146
PT15	X	115.4026	115.4384
	Z	-68.7372	-68.8611
PT16	X	116.7031	116.7395
	Z	-67.9864	-68.1081

The measuring precision of the globoidal cam intermittent section is pretty high, the tolerance between each point is rather small, and it proves that the measuring method of the equidistant model is suitable for the intermittent section. The reason why there is slightly larger deviation between X coordinate and the Z coordinate is that the profile cutting amount is pretty large when the cam is processed, and there is a slight gap between the globoidal cam and the roller, causing the phenomenon of 'jitter '.

Taking TC40 type globoidal cam as an example, measuring the eight characteristic lines (4 in intermittent section and 4 in indexing section). The calculation results of line profile tolerance of each characteristic line are shown in Table4.

Table4. Calculation results of line profile tolerance

line profile tolerance in intermittent section		
characteristic line	D_{tmax}	D_{tmin}
L1	0.043	-0.014
L2	0.038	-0.021
L3	0.041	-0.016
L4	0.035	-0.019
L5	0.213	-0.339
L6	0.014	-0.552
L7	0.221	-0.281
L8	0.436	-0.267

When the globoidal cam indexing mechanism is working in the intermittent section, the driven turnplate will stop from moving, and the exchange cutter manipulator will carry out the operation of grasping cutter and setting cutter. In order to prevent the tools from falling, the movement should be stable. The pretty high machining accuracy of globoidal cam profile can increase the density of measurement and improve the accuracy of tolerance detection. When working in the indexing section, the globoidal cam drives the driven turnplate to do rotary movement, and the driven turnplate drives the exchange cutter manipulator to finish the cutter exchange operation, in this process, we need to ensure the rotation precision and stability, so that make the manipulator working precisely in place. In addition, the phenomenon of vibration and entrapment should be avoided. In the indexing section, the change of curvature is large, the processing is difficult, and the contact measurement of cam profile brings about pretty big difficulty, the profile tolerance is also relatively large, so the measurement density can be reduced appropriately.

5 Summary

This paper researches on the influence caused by error of globoidal cam special machine tool on the error of globoidal cam profile, the machining error of special machine tool and coordinate transformation matrix is analyzed by using multi-body system theory, at the same time, the error model of the globoidal indexing cam profile is established, which lays certain foundation for the integration of design, processing and measurement to globoidal cam mechanism. The main conclusions are as follows:

1) According to the structure of globoidal cam special machine tool, establishing the machine tool error

transformation matrix based on multi-body system theory, and the mathematical model of globoidal cam profile error is established according to the generating machining theory of globoidal cam.

2) The establishment of mathematical model of globoidal cam profile machining error lays the theoretical basis for precision design and error distribution of special machine tool for globoidal cam profile, the quality detection and error evaluation of globoidal cam profile, and the tracing for globoidal cam machining error.

3) According to the method of generalizing machining and the equidistant model principle, establishing the theoretical profile equation of globoidal cam, the equidistant model equation and the axial sections characteristic line equation, which are used to evaluate the tolerance of the globoidal cam profile.

Putting forward the tolerance evaluation method of profile of the globoidal cam equidistant model, and then verifying it through experiment, experimental results are analyzed and the influence caused by globoidal cam profile tolerance on ATC transmission is discussed. This establishes the theoretical foundation for further separation and tracing of machining tolerance, establishment of the evaluation system and measuring standard about globoidal cam profile tolerance, and achieving the integration of design, machining and measurement.

Acknowledgments

This research was funded by the National Science and Technology Major Project, China, under contract 2013ZX04008-021. The authors would like to acknowledge the support and contribution from the Beijing No. 1 Machine Tool Plant for their help with manufacturing and be grateful to the Key Lab of Mechanical Engineering, Beijing University of Technology, China.

References

1. Han Yuemei. Research progress of automatic tool changer in machining center[J]. Equipment Manufacturing Technology, 2010, (5): 128-129.
2. Zhang Gaofeng, Yang Shiping, Chen Huazhang etc. Research and prospect of globoidal indexing CAM mechanism [J]. Mechanical Drive, 2003, 27(3): 1-5.
3. He Yi, Liu Yansong, Wang Tao etc. The review and prospect of arc indexing CAM mechanism research[J]. Light Industrial Machinery, 2003, (4): 7-10.
4. BU Fan-hua. Study on machining technology of globoidal cam employed in automatic tool changer (ATC) of machine center [D]. Beijing:College of Mechanical Engineering and Applied Electronics Technology, Beijing University of Technology, 2012.
5. W H Wang, C H Tseng, C B Tsay. Surface Contact Analysis for a Spatial Cam Mechanism[J]. Transaction of the ASME Journal of Mechanical Design, 1997, 119: 169-177.
6. H Y Cheng. Optimum Tolerance for Globoidal Cam Mechanisms [J]. JSME International Journal, Series C. 2002, 45(2): 519-526.
7. Yin M F, Lyu C Y, Zhao Z H. The theoretical study of CAM processing theory and the calculation method of the influence of CAM body deviation error on the profile error are studied[J]. Journal of shandong university of technology (natural science edition), 2001, 15(3):18-21.
8. Ji S T, Zhang Y M, Zhao J. The influence of center distance error of machining center of globoidalcam on convex contour error[J]. Journal of Beijing University of Technology, 2014, 40(6): 825-830.
9. Zhang Y M， Ji S T. Zhao J. Tolerance Analysis and Allocation of Special Machine Tool for Manufacturing Globoidal Cams [J]. International Journal of Advanced Manufacturing Technology, 2016,87:1597-1607.
10. Bu F H, Zhang Y M,Shang D G. Study on Machining Error of Globoidal Cam Profile Resulting from Rotational Deviation of Location of Part in Machining[C]].Advanced Materials Research 2012, 452-453:211-218.
11. WANG Hai-yan, ZHAO Ru-jia, LIU Chang-qi. Precision measurement and error evaluation of globoidal cam [J].Measurement Technique, 1999, 6:9-12.
12. WANG Xiao-fei, TANG Lin. Measurement theory and implement of globoidal indexing cam using cmm [J]. Journal of Tianjin University, 2011,44(11): 1024-1028.
13. P D Lin, J F Hsieh. Dimension inspection of spatial cams by cnc coordinate measuring machines[J]. Transaction of the ASME Journal of Manufacturing Science and Engineering, 2000, 122:149-157.
14. HU Bing-chen, KONG Qing-yi, XU Lian-tang. The discussion of precision index system of globoidal cam [J]. Transactions of the Chinese Society for Agricultural Machinery, 1994,25:13-16.
15. SONG Li-juan. Measurement of three-coordinate machine for globoidal indexing CAM[D].Dalian: College of Mechanical Engineering, Dalian University of Technology, 2005.
16. LIN Xiao-jun, WANG Zeng-qiang, SHAN Chen-wei. Method of probe radius compensation for free

surface measurement by CMM [J]. Aeronautical Manufacturing Technology, 2011, 10:75-78.

Analysis of Bridge-type Distributed-compliance Mechanism

Lei-Jie Lai [1] and Xiao-Qia Yin [2]

1College of Mechanical Engineering, Shanghai University of Engineering Science, Shanghai, 201620, China
[2]School of Electronic, Information and Electrical Engineering, Shanghai Jiao Tong University, Shanghai, 200240, China

Abstract. This paper analyses a class of bridge-type distributed-compliance mechanism, which has better performances than traditional bridge-type mechanisms using notch flexure hinges. An analytical model for the displacement amplification ratio and input stiffness calculations of the bridge-type mechanism is established based on the stiffness matrix method. The finite element analysis results are then given to validate the correctness of the analytical model. The differences of the analytical results with respect to the finite element analysis results are less than 8%, which demonstrate the high accuracy of the analytical model. The influences of the geometric parameters on the amplification ratio and input stiffness of the mechanism are also investigated using the analytical model to provide theoretical guidelines for the practical design.

1 Introduction

Recently, the piezoelectric actuator (PZT) is widely adopted due to its advantages of ultra-high resolution, fast response, high stiffness, compact size and high push force [1]. However, the relative small displacement produced by the PZT, which is about 10 μm/cm limits its applications of the larger working ranges. Therefore, in order to achieve larger output displacement, displacement amplification mechanisms are needed to amplify the stroke of PZT. Compared with the traditional lever-type mechanism, bridge-type flexure mechanism has the advantages of large amplification ratio, high resonant frequency and compact size, which becomes more and more popular.

In the previous literatures, many studies have been conducted to design and investigate bridge-type flexure mechanism. Lobontiu and Garcia formulated an analytical model for displacement and stiffness calculations of planar compliant mechanisms with single-axis flexure hinges based on the Castigliano's second theorem [3]. Ma et al. derived the ideal displacement amplification ratio of bridge-type mechanism using kinematic theory, and also derived a theoretic displacement amplification ratio using the elastic beam theory and work-energy theorem [4]. Some other methods developed for the analytical models of the bridge-type mechanisms are also reported.

All of the bridge-type compliant mechanisms mentioned in the previous literatures are lumped-compliance mechanisms, which use the notch flexure hinges. The existences of rigid bodies in the four arms of the lumped-compliance bridge-type mechanism have great weakness for the resonant frequency of the mechanism. Compared with the lumped-compliance mechanism, the distributed-compliance mechanism adopts four beam flexures to replace the notch flexure hinges and the rigid bodies, resulting in increased resonant frequency of the mechanism [5]. Moreover, the bending behavior of the beam flexure under the piezoelectric actuation allows for an acceptable distribution of the stresses in the distributed-compliance mechanism, which ensures a longer life time and better performances than traditional mechanical amplifiers in the dynamic applications. However, the mechanical properties of the bridge-type distributed-compliance mechanism still not been fully discovered. In the design stage, simple yet accurate enough analytical models are required to predict the performances of the mechanism and to further determine its architectural parameters according to the performance requirements.

In this paper, the mechanical properties and deformation characteristics of bridge-type distributed-compliance mechanism are deeply analyzed. An analytical model for the predictions of theoretical displacement amplification ratio and input stiffness is derived using the stiffness matrix method and validated by FEA. The comparisons between the analytical model and FEA results demonstrate the high accuracy of the analytical model. The influences of geometric and material parameters on the performances of the mechanism including the displacement amplification ratio and input stiffness are also well analyzed according to the analytical model.

2 Analytical model

The schematic of the bridge-type distributed-compliance mechanism with rhombic shell is depicted in Figure 1. The compliant mechanism mainly consists of four beam flexures that are connected in the way of bridge type.

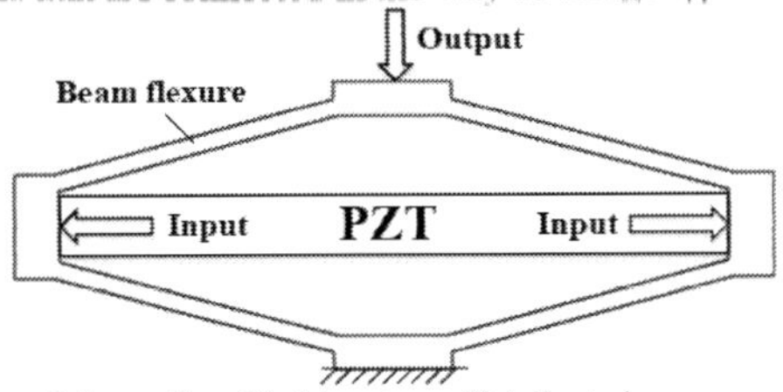

Figure 1. Schematic of bridge-type distributed-compliance

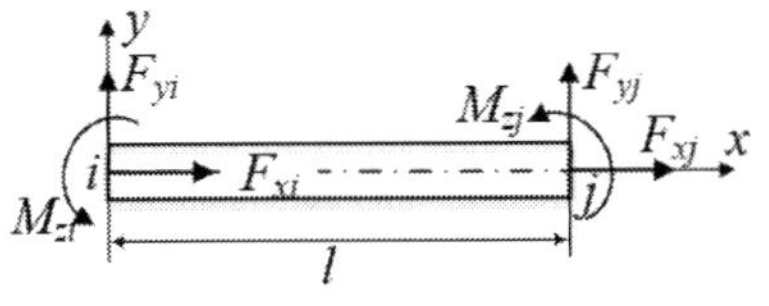

Figure 2. Model of a beam flexure mechanism

2.1 Model of single beam flexure.

The compliant mechanism contains four beam flexures, therefore, the stiffness characteristic of single beam flexure should be obtained first. As shown in Figure 2, two ends of the beam flexure are subjected to axial forces, bending moments and shear forces. In order to obtain the load-displacement relationship for the beam flexure, the right end of the beam is first fixed as shown in Figure 3(a). According to the elastic beam theory, the displacements of point i is

$$u_i = \frac{F_{xi} l}{EA},\ w_i = \frac{F_{yi} l^3}{3EI} - \frac{M_{zi} l^2}{2EI},\qquad \theta_i = -\frac{F_{yi} l^2}{2EI} + \frac{M_{zi} l}{EI} \tag{1}$$

where E is the Young's modulus of the material. A and I denote respectively the cross-sectional area and the second moment of inertia of the cross section about neutral axis. From the above three equations, the forces applied to point i can be calculated as

$$F_{xi} = \frac{EA}{l} u_i,\ F_{yi} = \frac{12EI}{l^3} w_i + \frac{6EI}{l^2}\theta_i,\qquad M_{zi} = \frac{6EI}{l^2} w_i + \frac{4EI}{l}\theta_i \tag{2}$$

Based on the force equilibrium theory, the forces applied to point j are

$$F_{xj} = -\frac{EA}{l} u_i,\ F_{yj} = -\frac{12EI}{l^3} w_i - \frac{6EI}{l^2}\theta_i,\qquad M_{zj} = \frac{6EI}{l^2} w_i + \frac{2EI}{l}\theta_i \tag{3}$$

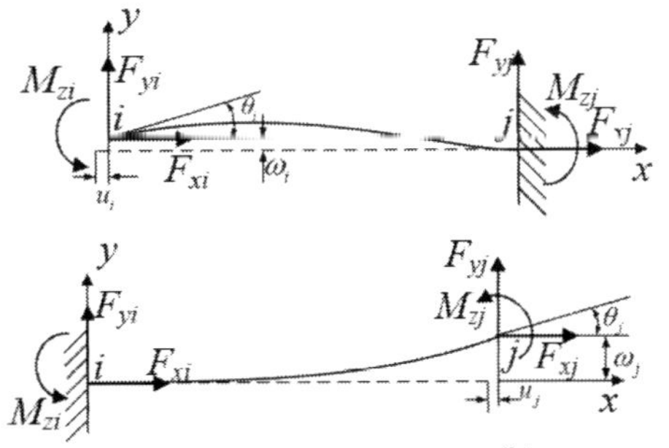

Figure 3. Two cases of beam structure. (a) Free-fixed. (b) Fixed-free

A similar procedure is applied by fixing the left end i of the beam, as shown in Figure 3(b). In this case, the forces applied to both ends can be calculated as

$$\begin{aligned} F_{xi} &= -\frac{EA}{l} u_j,\ F_{yi} = -\frac{12EI}{l^3} w_j + \frac{6EI}{l^2}\theta_j, \\ M_{zi} &= -\frac{6EI}{l^2} w_j + \frac{2EI}{l}\theta_j \\ F_{xj} &= \frac{EA}{l} u_j,\ F_{yj} = \frac{12EI}{l^3} w_j - \frac{6EI}{l^2}\theta_j, \\ M_{zj} &= -\frac{6EI}{l^2} w_j + \frac{4EI}{l}\theta_j \end{aligned} \tag{4}$$

By superimposing the results of the above two cases, as shown in Figure 2, the stiffness equations can be written as

$$\begin{aligned} F_{xi} &= \frac{EA}{l} u_i - \frac{EA}{l} u_j, \\ F_{yi} &= \frac{12EI}{l^3} w_i + \frac{6EI}{l^2}\theta_i - \frac{12EI}{l^3} w_j + \frac{6EI}{l^2}\theta_j \\ M_{zi} &= \frac{6EI}{l^2} w_i + \frac{4EI}{l}\theta_i - \frac{6EI}{l^2} w_j + \frac{2EI}{l}\theta_j \\ F_{xj} &= -\frac{EA}{l} u_i + \frac{EA}{l} u_j, \\ F_{yj} &= -\frac{12EI}{l^3} w_i - \frac{6EI}{l^2}\theta_i + \frac{12EI}{l^3} w_j - \frac{6EI}{l^2}\theta_j \\ M_{zj} &= \frac{6EI}{l^2} w_i + \frac{2EI}{l}\theta_i - \frac{6EI}{l^2} w_j + \frac{4EI}{l}\theta_j \end{aligned} \tag{5}$$

2.2 Displacement amplification ratio and input stiffness.

Because of the symmetrical structure of the bridge-type mechanism, only a quarter model of the mechanism shown in Figure 4 needs to be analyzed.

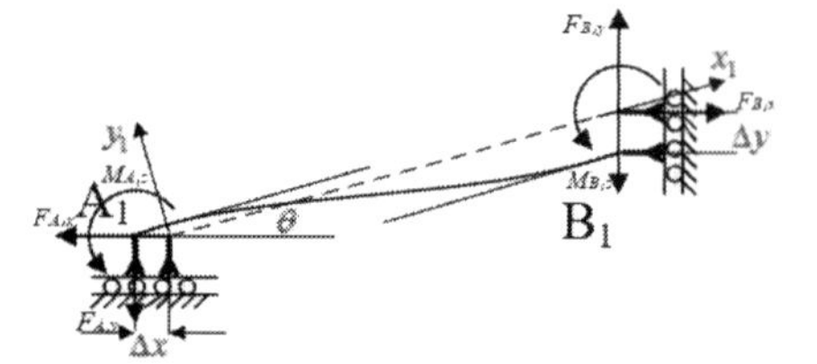

Figure 4. The quarter model of the bridge-type mechanism.

Because of the symmetrical structure of the mechanism, the orientations of the two rigid bodies are unchanged during the deformation process of the mechanism. According to the Euler-Bernoulli beam theory, the cross-sectional plane of deformed beam is still perpendicular to the deformed axis, the variations of rotation angles at Point A_1 and Point B_1 are both equal to zero. Therefore, the displacements at the Point A_1 and Point B_1 in the x_1- and y_1- directions can be obtained:

$$u_{A_1}=-\Delta x\cos\theta,\ w_{A_1}=\Delta x\sin\theta,\ \theta_{A_1}=0,$$
$$u_{B_1}=-\Delta y\sin\theta,\ w_{B_1}=-\Delta y\cos\theta,\ \theta_{B_1}=0 \tag{6}$$

The forces applied to Point A_1 and Point B_1 in the x_1- and y_1- directions can be calculated as:

$$F_{Ax_1}=-F_{Ax}\cos\theta-F_{Ay}\sin\theta$$
$$F_{Ay_1}=F_{Ax}\sin\theta-F_{Ay}\cos\theta \tag{7}$$

Then, the displacement amplification ratio and input stiffness can be calculated. The displacement amplification ratio and input stiffness without the consideration of the external force is

$$R_{amp}=\frac{\sin\theta\cos\theta\left(l^2-t^2\right)}{\left(t^2\cos^2\theta+l^2\sin^2\theta\right)},$$
$$K_{in}=\frac{2Ebt^3}{l\left(t^2\cos^2\theta+l^2\sin^2\theta\right)} \tag{8}$$

3 Finite element analysis validation

In this section, the finite element analysis (FEA) is performed to validate the analytical model formulated for evaluating the performances of this compliant mechanism. Three samples with different material and geometrical parameters and applied force conditions were set for the comparisons with FEA results, as listed in Table 1. The displacement amplification ratio and input stiffness calculated from the analytical model and the FEA model are listed in Table 2. The relative differences between the analytical results and FEA results of the three samples are all less than 8%, which indicates that the two results are in good agreement.

Table 1: Parameters of different samples.

Sample	E(GPa)	v	l(mm)	b(mm)	t(mm)	θ(°)	F_I (N)	F_O (N)
1	71	0.33	48	10	1.8	12.5	100	20
2	71	0.33	40	8	2	9	250	30
3	200	0.3	50	12	2.2	8	200	50

Table 2: Comparisons between the FEA and analytical results.

Sample		Analytical model	FEA model	Difference (%)
1	R_{amp}	3.39	3.41	0.6
	K_{in} (N/μm)	12.51	11.88	5.3
2	R_{amp}	4.39	4.36	0.7
	K_{in} (N/μm)	16.87	15.61	8
3	R_{amp}	8.20	8.08	1.48
	K_{in} (N/μm)	31.16	31.4	0.8

4 Influences of geometric parameters

The analytical model is established to analyze the influences of the geometric parameters on the displacement amplification ratio and input stiffness. l, b, t and θ are chosen as the designed geometric parameters of the mechanism. Figures 5-8 show the geometric parameters' influences on the displacement amplification ratio and input stiffness. It can be seen that the four geometric parameters have significant effects on the displacement amplification ratio and input stiffness:

b has no influence on the amplification ratio, which increases slightly while l increases and t decreases. The variation rate of Ramp decreases gradually with the increase of l.

The amplification ratio Ramp has a maximum value when θ reaches a threshold value. When θ decreases continuously, the amplification ratio Ramp decreases sharply.

The input stiffness K_{in} decreases with both l and θ increasing. Both of the two curves decrease sharply first and the gently. K_{in} also increases linearly with the increase of b.

https://doi.org/10.1051/matecconf/201821301005

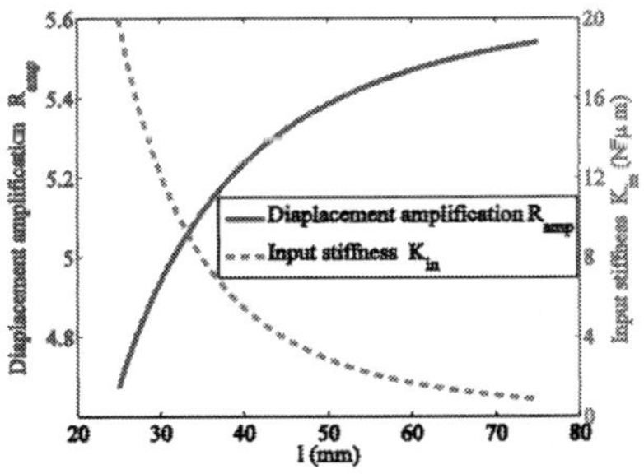

Figure 5: Influence of l on amplifier's performance.

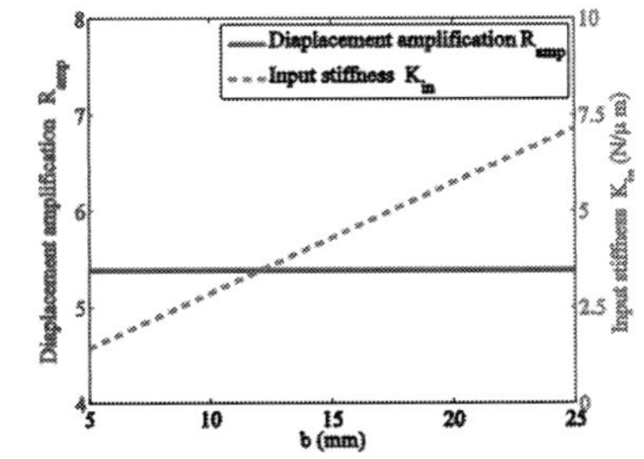

Figure 6: Influence of b on amplifier's performance.

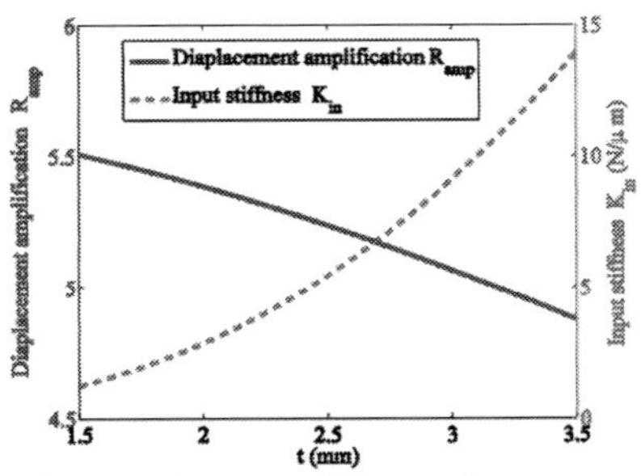

Figure 7: Influence of t on amplifier' performance.

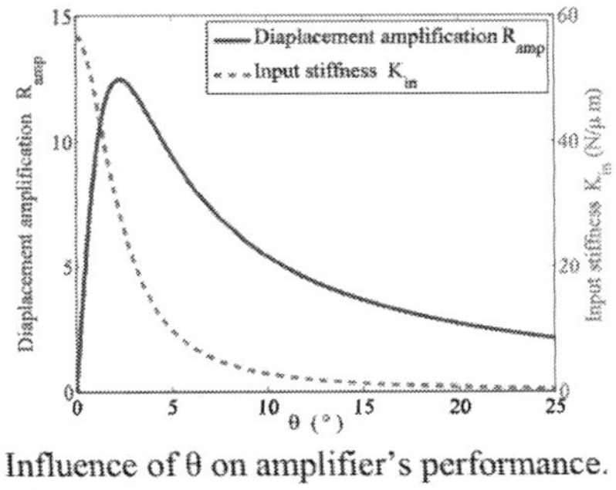

Figure 8: Influence of θ on amplifier's performance.

5. Conclusions

In this paper, a class of bridge-type distributed-compliance mechanism is analyzed. Due to distributed stresses and low mass, the bridge-type distributed-compliance mechanism has a longer life time and better performances than traditional mechanical amplifiers based on lever arm and flexure pivot. The stiffness matrix method is employed to derive analytical model to predict the displacement amplification ratio and input stiffness of the bridge-type mechanism. In order to verify the analytical model, the FEA simulations of the bridge-type mechanism are performed via the ANSYS Workbench. The comparisons show that the FEA and analytical results agree well with each other with the deviation of less than 8%. The relationships between the dimensions of the bridge-type mechanism and its performances including the displacement amplification ratio and input stiffness are represented graphically, which provide a theoretical basis for the practical design.

References

1. Kim, Dongmin, et al. Optimal design of a flexure hinge-based XYZ atomic force microscopy scanner for minimizing Abbe errors. *Review of Scientific Instruments* **76**.7 (2005): 073706 – 073706 -7.
2. Choi, Kee Bong, J. J. Lee, and S. Hata. A piezo-driven compliant stage with double mechanical amplification mechanisms arranged in parallel. *Sensors & Actuators A Physical* **161**. 1 (2010): 173-181.
3. Lobontiu, Nicolae, and E. Garcia. Analytical model of displacement amplification and stiffness optimization for a class of flexure-based compliant mechanisms. *Computers & Structures* **81**. 32 (2003): 2797-2810.
4. Ma, Hong Wen, et al. Analysis of the displacement amplification ratio of bridge-type flexure hinge. *Sensors & Actuators A Physical* **132**.2(2006):730-736.
5. Polit, Sebastian, and J. Dong. Development of a High-Bandwidth XY Nanopositioning Stage for High-Rate Micro-/Nanomanufacturing. *IEEE/ASME Transactions on Mechatronics* **16**.4 (2011): 724-733.

Accuracy Simulation Method Based on Assembly Process Information Mapping

Jianjun Tang [1] and Haitao Liu [1]

[1] *AVIC Chengdu Aircraft Industry (Group) Co., Ltd., Chengdu, 610092, P.R. China*

Abstract. Assembly process information has a direct impact on product manufacturing accuracy. In order to accurately predict product accuracy, a simulation method based on assembly process information mapping was proposed. The assembly deviation information was expressed as the in-part deviation reference, feature deviation Vector direction, feature deviation vector size, deviation transmission connection form between the parts and the deviation transmission direction, and so on. The mapping between the assembly process information and the deviation transmission information was established. Then the assembly accuracy simulation algorithm including the main algorithm and assembly deviation transfer model building algorithm was established. Taking C919 internal flap assembly as an example, the effectiveness of this method had been verified.

1 Introduction

Aircraft assembly accuracy is an important factor affecting the performance of the whole machine. Due to the different assembly constraints between the parts, the dimensions deviation of the parts has a different degree of influence on the overall assembly accuracy [1]. Because the assembly process defines the assembly sequence of parts and assembly constraints, so In order to obtain more reliable simulation results of assembly accuracy, the information of assembly process needs to be fully integrated.

The assembly process defines all the information that can reflect the deviation transfer path and directly affect the formation of assembly accuracy [2-4]. Literature [5-7] studied the assembly process information model and the tolerance information model, but lacks the research on the relationship between the two.

In order to analyze the influence of assembly process information on the assembly accuracy comprehensively and obtain a more reliable prediction value of assembly accuracy, a simulation method based on assembly process information mapping is proposed. Through the construction of the mapping between multi-factor assembly process information and assembly deviation transfer information, as well as the assembly accuracy simulation algorithm to achieve a reliable assembly accuracy simulation.

2 Multi-factor assembly process information definition

Improper tolerance allocation or unreasonable process planning will lead to over-accuracy of assembly key characteristics. The existing aircraft tolerance distribution is mainly based on experience and knowledge. There is no systematic analysis of the factors influencing the accuracy of the assembly key features in the assembly process. It is difficult to find out the accuracy of the assembly key characteristics before the aircraft enters the assembly production.

According to the characteristics of aircraft assembly process planning, assembly process consists of multiple assembly processes. By extending the assembly process model, the assembly process information related to assembly deviation transmission is analyzed, which mainly includes product model information, process design information and tooling model information. Product model information describes product design dimensions and tolerance; process design information describes assembly sequences, assembly features, assembly constraints, and positioning references; tooling model information describes tooling dimensions and tolerances.

3 Mapping relationship building

In order to establish a more realistic transfer path of assembly deviation, it is necessary to determine the mapping relationship between multi-factor process information and deviation transfer information.

3.1 Abstract Expression of Assembly Deviation Transfer Information

Assembly deviation transmission refers to the assembly deviation along the direction of deviation transmission to reach the key characteristics of the assembly process nodes. Assembly deviations are propagated once at each deviation transfer node. Deviations are mainly composed of two parts: the own deviation of the node and the variation deviation of the previous node. Deviation can be expressed as

$$T_{i+1} = a_i \cdot T_i + t_{i+1} . \quad (1)$$

Where:

T_i said the previous node's deviation value;

a_i represents the coefficient of variation transmitted by T_i to the next node, which is determined by the deviation transfer path;

t_{i+1} indicates the own deviation value of this node;

T_{i+1} indicates the deviation value formed by this node.

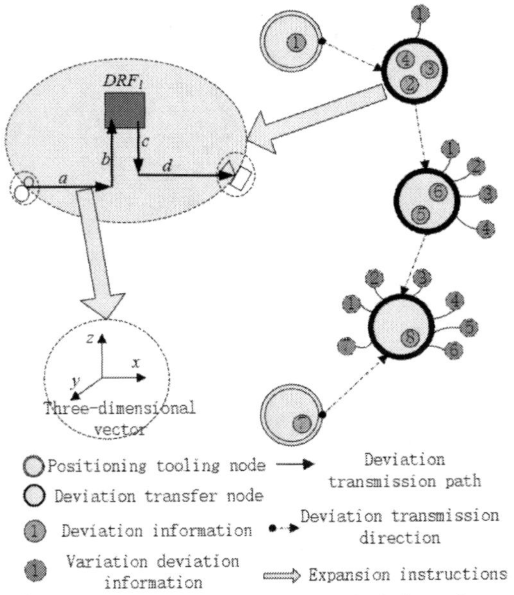

Fig. 1 Assembly deviation transfer information

Assembly deviation transfer information abstract expression as shown in Fig. 1. Aircraft assembly process is generally located by the tooling, so the initial node of assembly deviation transmission is positioning tooling node. The assembly deviation transfer process mainly passes two types of deviations: the deviations introduced by this node and the deviations introduced before this node. Within each deviation transfer node, the deviation is transmitted in a certain deviation transfer path, which must go through the in-part deviation reference, which is generally the part positioning reference. The path of deviation transmission is determined by the design dimensional information of the part, which mainly contains the information of three-dimensional vector size and direction.

Assembly deviation transfer information is affected by product design information and process information. And they have corresponding relationships. By discussing the mapping relationship between the assembly process information and the deviation transfer information, a more reliable deviation transfer path can be obtained, which is the basis of the assembly accuracy simulation.

3.2 The Mapping Relationship for Assembly Deviation Transfer

According to the analysis of pheromone information of assembly deviation transmission, the assembly deviation transmission model mainly includes five types of information: in-component deviation reference standard, characteristic deviation vector direction, characteristic deviation vector size, deviation transmission connection form between parts and deviation transmission direction. These five types of information have one-to-one correspondence with the positioning datum, assembly characteristic, design tolerance, assembly constraint and assembly sequence in MBOM, as shown in Fig. 2.

4 Assembly accuracy simulation based on Monte Carlo

Through the assembly accuracy simulation can fully analyze the assembly process on the accuracy of key features. Before actual assembly and assembly, the input condition of assembly accuracy simulation is the design tolerance zone. How to divide the design tolerance zone into a set of effective deviation values is a prerequisite for assembly simulation. The use of Monte Carlo algorithm can effectively solve the above problems.

The simulation of assembly accuracy based on Monte Carlo algorithm is shown in Fig. 3. The main work includes building the assembly deviation transfer model and setting the correlation value of the Monte Carlo algorithm. The main steps of the algorithm are as follows:

Step 1: Determine the key features that need to be simulated, using this as a basis to resolve the associated assembly units, thereby reducing the amount of computation;

Step 2: Determine if the corresponding assembly deviation transfer model has been constructed

If Yes Jumping to step 3 to continue execution;

If No Jumping to assembly deviation transfer model building algorithm;

Step 3: Define the number of deviation samples n, setting n = 5000;

Step 4: Enter the tolerance field for all relevant features, and use the random number generation principle to generate a deviation random sample matrix of key features;

Step 5: Calculate the distribution of the simulation results and determine whether the deviation within ± 3 σ is out of tolerance;

If No Then come to meet the accuracy requirements of the results;

If Yes The assembly process need to be optimized.

Assembly deviation transfer model building algorithm specific steps are as follows:

Step 1: Analyze the assembly constraints, the associated assembly features and positioning features in the corresponding assembly unit;

Step 2: Build the deviation reference base on the location feature;

Step 3: The characteristics and the direction of the deviation transmission are expressed by vectors;

Step 4: In the corresponding assembly unit, according to the assembly constraints to build the relationship between the parts of the deviation transfer connection.

Step 5: Return to the main algorithm.

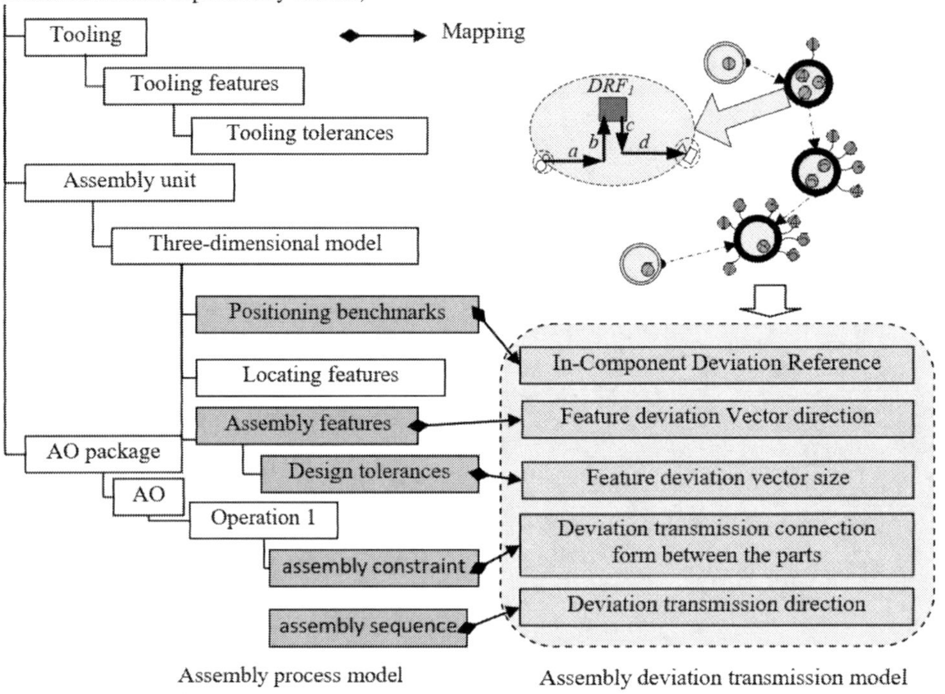

Fig. 2 The mapping between assembly process information and deviation transmission information

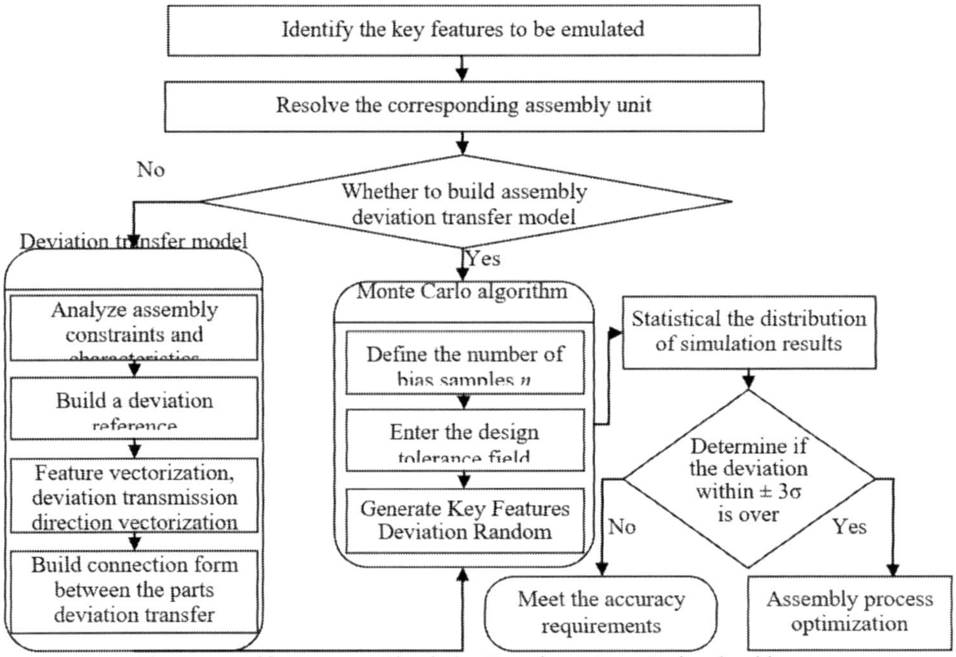

Fig. 3 Assembly Accuracy Simulation Based on Monte Carlo Algorithm

https://doi.org/10.1051/matecconf/201821301006

5 Examples of verification

Taking the C919 inner flap component assembly as an example, an assembly accuracy simulation process based on assembly process information mapping is described. The assembly constraints contained in the assembly unit of the inner flap components are analyzed; the corresponding assembly features and positioning features are constructed; According to the vector direction and size of the corresponding surface in the 3D model of the product, a corresponding assembly deviation transfer model is constructed, shown in Fig. 4.

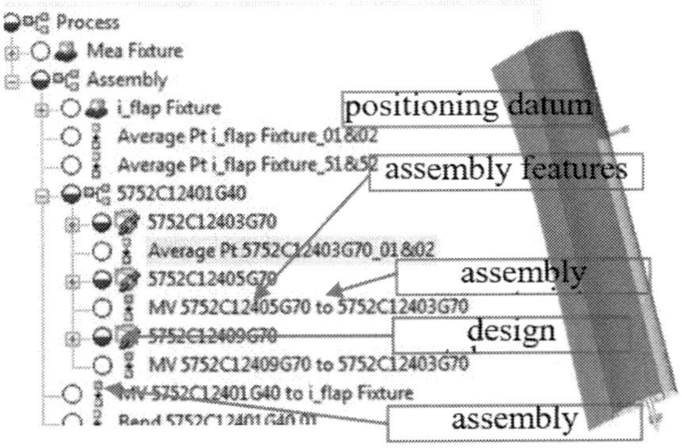

Fig. 4 Inner flap assembly deviation transfer model

The accuracy simulation results of the target key features included in the inner flap components are shown in Fig. 5.

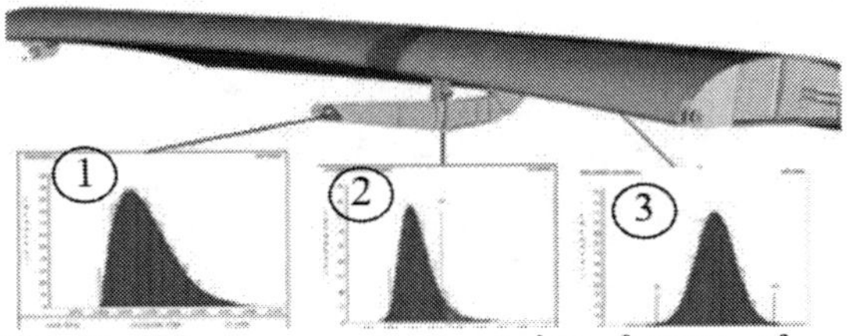

Fig. 5 Slab components target key features of precision simulation results

In Fig. 5, ①, ② and ③ represent the statistic results of the simulation results of the inner profile flaps. It can be seen from the simulation results in the figure that there is a possibility that the axis deviation of the 2 # pulley carrier tab joint hole may be out of tolerance. By optimizing the assembly process, assembly precision can be effectively controlled.

6 Summary

Through establishing the mapping relationship between assembly process information and assembly deviation transfer information, a real assembly deviation transfer model is constructed. The method combines the effects of positioning datum, assembly features, design tolerances, assembly constraints, and assembly sequences on assembly deviation transfer. Using the assembly precision simulation based on Monte Carlo algorithm, the accurate prediction of the assembly key characteristics can be realized.

References

1. Jin, J. (2005). Process-oriented tolerancing for multi-station assembly systems. Iie Transactions, 37(6), 493-508.
2. Xu, S., & Keyser, J. (2014). Geometric computation and optimization on tolerance dimensioning. Butterworth-Heinemann.
3. Samuel H. Huang, Qiu Liu, & Rami Musa. (2004). Tolerance-based process plan evaluation using monte carlo simulation. International Journal of Production Research, 42(23), 4871-4891.
4. Hong, Y. S., & Chang, T. C. (2002). A comprehensive review of tolerancing research. International Journal of Production Research, 40(11), 2425-2459.
5. Bing, Y., & Zhang, L. X. (2006). Introduction of integrated assembly model for assembly process planning. Journal of System Simulation.
6. Liu, Y. (2003). Hierachical representation model and its realization of tolerance based on feature. Chinese Journal of Mechanical Engineering, 39(3), 649-50.
7. Guilford, J., & Turner, J. (1993). Representational primitives for geometric tolerancing. Computer-Aided Design, 25(9), 577-586.

https://doi.org/10.1051/matecconf/201821301007

Investigation on influence of cutting parameters on spindle vibration of CNC wood milling machine

Van Thuy Tran [1] **and Huu Loc Nguyen** [2]

[1] *Faculty of Mechanical Engineering, HCMC University of Technology, VNU-HCM, Viet Nam (PhD. Student)*
[2] *Faculty of Mechanical Engineering, HCMC University of Technology, VNU-HCM, Viet Nam*

Abstract. In machining operation, the cutting parameters greatly influences on the spindle vibration of a CNC wood milling machine. The paper presents the effect of the cutting parameters such as feed rate, cutting speed, and cutting depth on the vibration amplitude of the spindle when machined on CNC milling machine using Box-Hunter method of experimentation. The lowest natural frequency of this machine is 250 Hz. Experimental results have established a second-order regression equation that demonstrates the effect of three parameters such as feed rate, cutting speed, and cutting depth on the vibration amplitude of the spindle. From that base, determine the most reasonable cutting parameters when machining on CNC wood milling machines so that the spindle vibration amplitudes is minimal. In addition, The comparison results show that the spindle head vibration amplitude of the machine using the bolt joint is larger than the spindle head vibration amplitude of the machine using the weld joint.

1 Introduction

Traditionally, the base and other major components of a machine tool have been made of gray or nodular cast iron, which has the advantages of low cost and good damping, but the disadvantage of heavy weight

In modern machine design, lightweight structures are desirable because of ease of transportation, higher natural frequencies, and lower inertial forces of moving members. Lightweight designs are a basic goal in machine tool design and require fabrication processes such as mechanical fastening (bolts and nuts) of individual components and welding. Bolted and welded machine tool structures consists of pre-cut stock materials such as plates, tubes, channels, and angles which are joined together to form the structure in Fig.1. Such stock items are available in a wide range of sizes and shapes and have some highly desirable mechanical properties such as formability, machinability and weldability. Of special interest are round and rectangular tubes whose closed cross sections have a very high stiffness-to-weight ratio [1].

Welded machine tool structures provide easy scalability in terms of size and outstanding flexibility in terms of fast design and fabrication; however, damping of the structure is a very critical issue. Unlike cast iron or polymer concrete-based components, welded steel plates have virtually no internal damping and are therefore prone to unwanted vibrations. Despite the shortcomings listed above, designing and building a machine as a welded structure has the big advantage of a much lighter design with a substantially shorter lead time compared to a cast design.

(a)

(b)

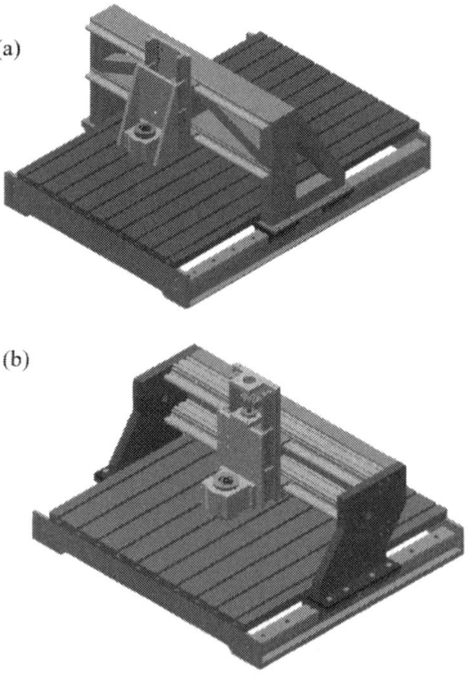

Figure 1. The CNC machine tool structures: (a) Welded structure and (b) Bolted structure

The vibrations are induced due to machine faults, cutting parameters, cutting tool, work piece deformation, etc. Cutting parameters especially feed rate, depth of cut and spindle speed if not properly controlled can increase the vibration in the machining process. Too high vibration is very dangerous and may lead to serious injury. Therefore, vibration must be minimized to stabilize the cutting process, improve surface roughness and prevent tool from breaking. The effect of cutting tool vibration on surface roughness of workpiece in dry turning operation and a correlation between surface roughness and cutting tool vibration in turning [2]. Forced and self-excited vibration limits the stability of cutting and lead to tool and machine damage. It also limits the productivity of machining operation.

Controlling spindle vibration of machine tool is one of the approaches for enhancing the quality of the product. Investigated the Effect of spindle forced vibrations on surface roughness, dimensional accuracy, and tool wear in vertical milling of AISI P20 [3]. Several researchers have studied the effect of cutting parameters such as spindle speed, feed rate, axial and radial depth of cut, appropriate selection of tool material and tool geometry in machining of aluminum based alloys [4, 5] Prediction of surface roughness in end face milling based on Gaussian process regression and cause analysis considering vibration of cutting tool [6]. Developed experimental plan and conducted experiments based on central composite design (CCD) of response surface methodology (RSM) [7]. Experimental investigation on the effect of cutting parameters on the spindle vibration and surface roughness in the precision end-milling process using the singular spectrum analysis [8-10]. An experimental investigation on the effect of machining and geometrical parameters such as spindle speed, feed rate, axial and radial depth of cut and radial rake angle on responses during end milling operation on the surface roughness and its defects of AL 6061-T6 [11]. Effect of tool vibration, cutting conditions on surface roughness during lathe dry turning process [12, 13]

Surface roughness is mainly affected by machine tool vibration. To improve the surface roughness, it is very essential to know the influences of cutting parameters on machine tool vibration. The purpose of this research is to investigate on the influences of cutting parameters such as feed rate, cutting speed and cutting depth on spindle vibration amplitude of welded structure of CNC wood milling machine structure using Box-Hunter quadratic experimental model.

2 Experimental methodology

2.1 Experimental setup

The experiments have been performed on welded structure of CNC wood milling machine. The lowest natural frequency of this machine is 250 Hz. The device used for the spindle vibration test of CNC milling machine consists of instrument accelerometer sensor PCB 603C01, PCB Group, the accelerometer sensitivity 10.2 mV/g. The Multi-Channel NI SCXI-1000DC System was used to analyze the vibration detected by the accelerometer. For a good operating of the NI SCXI-1000DC system, the NI Signal Express software displayed the result of vibration data and a personal computer. The workpiece material used in this study is fir wood. The experimental setup for operation is shown in Fig.2. In particular, the acceleration sensor is attached to the spindle head of the machine, the NI SCXI-1000DC measurement device is connected to the acceleration sensor and the computer. Measurement data is collected and displayed through the NI Signal Express software.

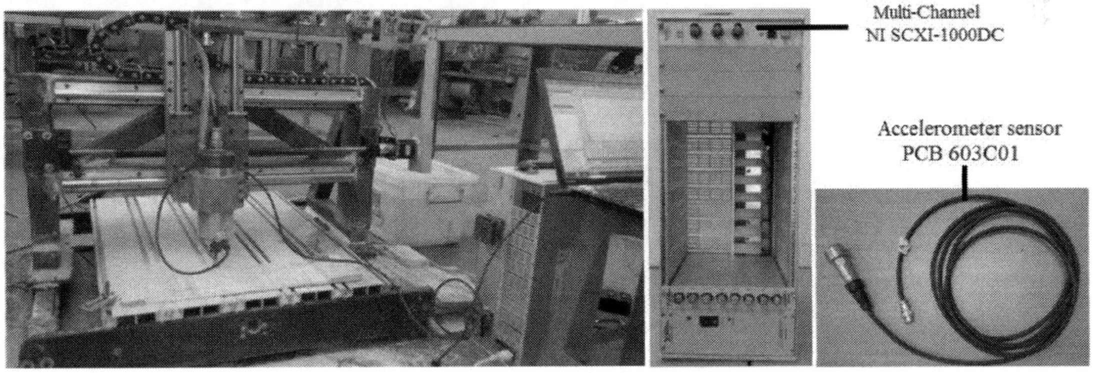

Figure 2. Experimental Setup

2.2 Design of Experiment (DOE)

The machining parameters were selected are spindle speed, feed rate and depth of cut. Cutting parameters were considered as input factors, spindle vibration were considered as output parameters during this analysis. The number of levels for each factors considered in this DOE is as shown in Table 1.

Experimental matrix based on the symmetrically mixed quadratic planning model with 3 coded factors as shown in Table 2.

Table 1. Design schema of cutting parameters and their levels

Variables or parameter	Factor symbol		interval of variation	Variable range of cutting parameters				
	Actual factor	Code factor		level - α	level (-1)	level (0)	level (+1)	level + α
Spindle Speed [rev/min]	n	x_1	5000	2590	6000	11000	16000	19410
Depth of Cut [mm]	t	x_2	0.6	0.29	0.7	1.3	1.9	2.31
Feed Rate [mm/ min]	s	x_3	500	159	500	1000	1500	1841

Table 2. Experimental matrix of 3 factors

	N°	x_0	x_1	x_2	x_3	x_1x_2	x_1x_3	x_2x_3	x_{11}	x_{22}	x_{33}
Number of experi-ments in nucleus 2^k	1	+1	+1	+1	+1	+1	+1	+1	+1	+1	+1
	2	+1	-1	+1	+1	-1	-1	+1	+1	+1	+1
	3	+1	+1	-1	+1	-1	+1	-1	+1	+1	+1
	4	+1	-1	-1	+1	+1	-1	-1	+1	+1	+1
	5	+1	+1	+1	-1	+1	-1	-1	+1	+1	+1
	6	+1	-1	+1	-1	-1	+1	-1	+1	+1	+1
	7	+1	+1	-1	-1	-1	-1	+1	+1	+1	+1
	8	+1	-1	-1	-1	+1	+1	+1	+1	+1	+1
Number of experi-ments in point start (*)	9	+1	+1.682	0	0	0	0	0	$2\sqrt{2}$	0	0
	10	+1	-1.682	0	0	0	0	0		0	0
	11	+1	0	+1.682	0	0	0	0	$2\sqrt{2}$	$2\sqrt{2}$	0
	12	+1	0	-1.682	0	0	0	0	0		0
	13	+1	0	0	+1.682	0	0	0	0	$2\sqrt{2}$	
	14	+1	0	0	-1.682	0	0	0	0	0	$2\sqrt{2}$
									0	0	$2\sqrt{2}$
Number of experi-ments in the planning center	15	+1	0	0	0	0	0	0	0	0	0
	16	+1	0	0	0	0	0	0	0	0	0
	17	+1	0	0	0	0	0	0	0	0	0
	18	+1	0	0	0	0	0	0	0	0	0
	19	+1	0	0	0	0	0	0	0	0	0
	20	+1	0	0	0	0	0	0	0	0	0

The mean results of measuring the spindle vibration amplitude of 20 experiments and each experiment were repeated 7 times as shown in Table 3.

Table 3. The results the spindle vibration amplitude

N⁰	$\bar{y}$	s_i^2	N⁰	$\bar{y}$	s_i^2
1	0.351	0.0004	11	0.398	0.0006
2	0.581	0.0084	12	0.301	0.0014
3	0.116	0.0007	13	0.224	0.0025
4	0.525	0.002	14	0.176	0.0004
5	0.31	0.0015	15	0.075	0.0002
6	0.489	0.0021	16	0.07	0.0002
7	0.09	0.0005	17	0.075	0.0001
8	0.491	0.0022	18	0.07	0.0001
9	0.278	0.0007	19	0.075	0.0001
10	0.844	0.0158	20	0.07	0.0001

3 Results and discussion

3.1 The coefficients of regression equation

With this Box-Hunter experimental model, the coefficients of regression equation were obtained by the results of experimental data and expressed by the following Eq. 1.

$$y = 0.0731 - 0.1597x_1 + 0.0498x_2 + 0.0208x_3 \\ + 0.0489x_1x_2 - 0.0086x_{13} + 0.0104x_{23} + \\ + 0.1689x_1^2 + 0.0943x_2^2 + 0.0413x_3^2 \quad (1)$$

3.2 The testing of the individual regression coefficients

After calculating the value of the coefficients of the regression equation, it is necessary to test their significance and eliminate the insignificant individual regression coefficients [14, 15]. The computational results are shown in Table 4.

Table 4. Significance test of individual regression coefficients

Tern	Value of Coefficient	SE coefficient	T	P-Value	
b_0	0.0731	0.01048	6.977	0.00	
b_1	-0.1597	0.006953	-22.861	0.00	
b_2	0.0498	0.006953	7.078	0.00	
b_3	0.0208	0.006953	3.183	0.086	Insignificant
b_{12}	0.0489	0.006769	5.517	0.00	
b_{13}	-0.0086	0.006769	-0.812	0.436	Insignificant
b_{23}	0.0104	0.006769	1.004	0.339	Insignificant
b_{11}	0.1689	0.009085	24.920	0.00	
b_{22}	0.0943	0.009085	13.873	0.00	
b_{33}	0.0413	0.009085	6.064	0.00	

In the Significance test table, P - value of less than 0.05 indicates that the model terms can be considered significant. Table 4 presented of Significance test table for regression equation coefficients. There are 3 coefficient b_3, b_{13} and b_{23} are not significant because its P value is more than 0.05. We remove these coefficients, then we rewrite the matrix and recalculate the coefficients. Through the backward elimination process, the final quadratic models of response equation in terms of coded factors are presented as follows Eq. 2.

$$y = 0.0731 - 0.1597x_1 + 0.0498x_2 + 0.0489x_1x_2 + \\ + 0.1689x_1^2 + 0.0943x_2^2 + 0.0413x_3^2 \quad (2)$$

3.3 Testing lack of fit in regression

For testing lack of fit in regression, to perform the analysis of variance for Significance of regression is shown in Table 5

Table 5. Analysis of variance for Significance of regression

Source	Sum of Squares	DF	Mean Square	F	P-Value
Model	0.8847	6	0.14746	68.94	0.00
Residual Error	0.0278	13	0.00214		
Lack-of-Fit	0.027767	8	0.003471	462.78	0.00
Pure Error	0.000038	5	0.000008		
Total	0.9125	19			

The model was developed with 95% confidence interval. Because $F_{model} = 68.94 > F_{0.05,6,13} = 2.92$ and $F_{LOF} = 462.78 > F_{0.05,8,5} = 4.82$. Also, note that the P-value for F_{model} and F_{LOF} (shown in Table 5) is considerably smaller than $\alpha = 0.05$. The regression model is fitting well.

In the actual form, the regression equation shows the relation between the feed rate (s), cutting speed (n), cutting depth (t) and the spindle vibration amplitude (A) of the CNC machine are presented as follows Eq. 3.

$$A = 1.993 - 20.212.10^{-5}n - 0.7807t - \\ - 3.304.10^{-4}s + 0.0166.10^{-3}nt + 0.0676.10^{-7}n^2 \quad (3) \\ + 0.262t^2 + 1.652.10^{-7}s^2$$

3.4 Evaluation of experimental results

In the milling process, the machining parameters adopting the spindle speed, the feed rate, and the depth of cut were chosen as the three factors so as to investigate the influences on the vibration amplitude. The influences of machining parameters on the performance of the vibration amplitude have been analyzed, which is based on the above proposed mathematical model.

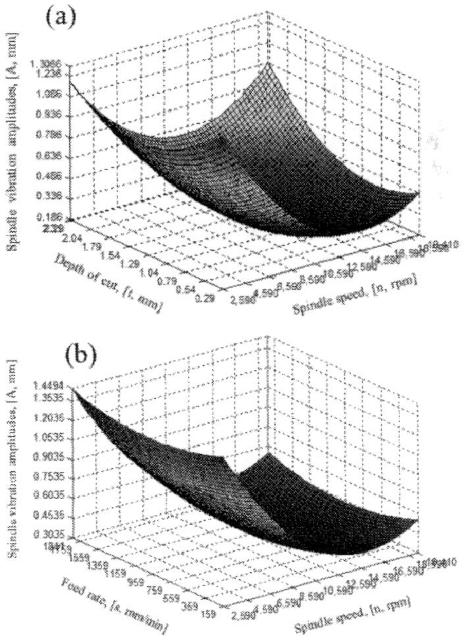

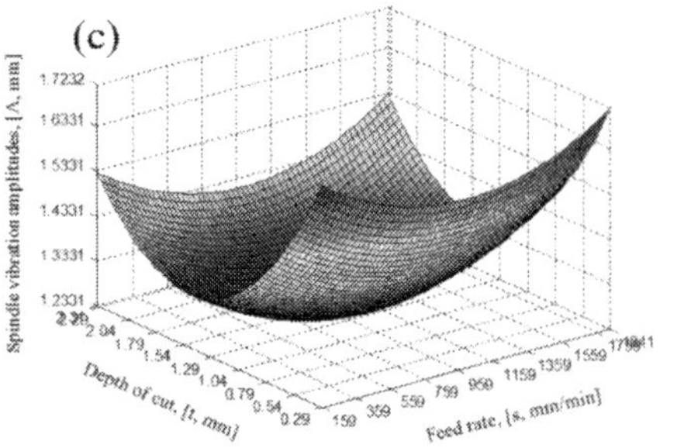

Figure 3. (a) Effect of n and t on the spindle vibration amplitude; (b) Effect of n and s on the spindle vibration amplitude and (c) Effect of t and s on the spindle vibration amplitude

In graphs Fig. 3 shows the relationship between the feed rate, cutting speed, cutting depth and the spindle vibration amplitude of the CNC milling machine. It can be seen that the tendency of spindle vibration amplitudes slightly decreases with an increase of spindle speed. Increasing the spindle speed promotes the rotation stability of spindle head, which results in suppressing the vibration amplitude. As seen above, the spindle vibration amplitudes obviously increase with increasing the feed rate.

In the cutting process, the resistance force acting on the cutting tool is mainly generated by the status of feed rate and depth of cut. The increase of cutting depth and feed rate causes the increase in the resistance force at the tooltip and cutting feed direction, which excites more vibration appearance. It can be obviously seen that the spindle vibration amplitudes increase with increasing the value of cutting depth and feed rate. As the results of analysis discussed above, the spindle vibration amplitudes are principally influenced by the spindle speed, feed rate and cutting depth.

In Fig. 3, the local reasonable cutting parameters valued (within the scope of the survey) so that the spindle vibration amplitudes is minimal as follows: n = 14280 rpm, s = 1043 mm/min and t = 1.043 mm => spindle vibration amplitude: A_{min} = 0.017 mm.

4 The machine structure using the bolt joint

Machine structure using a bolt joint and an experimental setup of the spindle head of the CNC milling machine is shown in Fig. 4.

With the same cutting parameters as above, the results of measuring the spindle vibration of 20 experiments and each experiment were repeated 7 times as shown in Table 6.

The regression equation shows the relation between the feed rate (s), cutting speed (n), cutting depth (t) and the spindle vibration amplitude (A) of the milling machine are presented as follows Eq. 4.

$$A = 1.993 - 20.212.10^{-5} n - 0.7807t - \\ - 3.304.10^{-4} s + 0.0166.10^{-3} nt + 0.0676.10^{-7} n^2 + \\ + 0.262t^2 + 1.652.10^{-7} s^2 \qquad (4)$$

Figure 4. Machine structure using a bolt joint and Experimental Setup

Table 6. Experimental results

N^0	$\bar{y}$	s_i^2	N^0	$\bar{y}$	s_i^2
1	0.355	0.0014	11	0.0097	0.0006
2	0.576	0.0146	12	0.0056	0.0014
3	0.121	0.0004	13	0.001	0.0025
4	0.576	0.0059	14	0.0016	0.0004
5	0.34	0.0006	15	0.0001	0.0002
6	0.541	0.0211	16	0.069	0.0001
7	0.098	0.0005	17	0.067	0.0001
8	0.575	0.0119	18	0.07	0.0001
9	0.223	0.0002	19	0.075	0.0001
10	0.862	0.0133	20	0.072	0.0001

The comparison results of spindle vibration amplitude for two different machine structure types are the structures using the welded joint and the bolt joint.The results show that the spindle head vibration amplitude of the machine using the bolt joint is larger than the spindle head vibration amplitude of the machine using the weld joint.

5 Conclusions

The paper presents the effect of the cutting parameters such as feed rate, cutting speed, and cutting depth on the vibration amplitude of the spindle when machined on CNC milling machine by experimental design method (DOE). Using the Box-Hunter method of experimentation to construct a second-order regression equation representing the relationship between the cutting parameters and the vibration amplitude of the spindle to survey the effect of the cutting parameters on the spindle vibration in machining process. The results show that as the spindle speed increases, the rotational stability of the spindle head is promoted, which reduces the vibrations amplitude and the spindle vibration amplitudes increase with increasing the value of cutting depth and feed rate to cause the unstability of cutting process and exhibit the result of the worst machined surface. In addition within the scope of survey also identified the most appropriate cutting parameters for machining on CNC wood milling machine.

The comparison results of spindle vibration amplitude for two different machine structure types are the structures using the welded joint and the bolt joint show that the spindle head vibration amplitude of the machine using the bolt joint is larger than the spindle head vibration amplitude of the machine using the weld joint.

Acknowledgement: This research is funded by Vietnam National University - Ho Chi Minh City (VNU-HCM) under grant number B2016-20-04.

References

1. E. Bamberg, Principles of rapid machine design, Massachusetts Institute of Technology, (2000).
2. S.Y. Kassab, Y.K. Khoshnaw, Engineering and technology **25** 879-889 (2007).
3. S. Zahoor, N.A. Mufti, M.Q. Saleem, M.P. Mughal, M.A.M. Qureshi, The International Journal of Advanced Manufacturing Technology **89** 3671-3679 (2017) .
4. M. Sayuti, A.A. Sarhan, M. Hamdi, The International Journal of Advanced Manufacturing Technology **67** 833-849 (2013).
5. G. Mahesh, S. Muthu, S. Devadasan, The International Journal of Advanced Manufacturing Technology **77** 369-381 (2015).
6. G. Zhang, J. Li, Y. Chen, Y. Huang, X. Shao, M. Li, The International Journal of Advanced Manufacturing Technology **75** 1357-1370 (2014).
7. S.S. Baraskar, S. Banwait, S. Laroiya, Materials and Manufacturing Processes **28** 348-354 (2013).
8. C.-C. Chen, N.-M. Liu, K.-T. Chiang, H.-L. Chen, The International Journal of Advanced Manufacturing Technology **63** 797-815 (2012).
9. B. Kilundu, P. Dehombreux, X. Chiementin, Mechanical Systems and Signal Processing **25** 400-415 (2011).
10. N. Kusuma, M. Agrawal, P. Shashikumar, (2014).
11. N. Zeelanbasha, V. Senthil, B. Sharon Sylvester, Metalurgija **57** 121-124 (2018) .
12. D. Dimla Sr, Journal of materials processing technology **155** 1708-1715 (2004).
13. M. Thomas, Y. Beauchamp, A. Youssef, J. Masounave, Computers & industrial engineering **31** 637-644 (1996).
14. D.C. Montgomery, *Design and analysis of experiments*, John wiley & sons, (2017).
15. H.L. Nguyen, *Experimental Design*, National University publisher HCMC, Viet Nam, (2014).

Determining Optimal Gear Ratios of a Two-stage Helical Reducer for Getting Minimal Acreage of Cross Section

Nguyen Khac Tuan[1], Vu Ngoc Pi[2], Nguyen Thi Hong Cam[2], Tran Thi Phuong Thao[2], Ho Ky Thanh[2], Le Xuan Hung[2], Hoang Thi Tham[3]

[1]*Automotive and Power Machinery Faculty, Thai Nguyen University of Technology, Thai Nguyen 23000, Vietnam*
[2]*Mechanical Engineering Faculty, Thai Nguyen University of Technology, Thai Nguyen 23000, Vietnam*
[3]*International Training Faculty, Thai Nguyen University of Technology, Thai Nguyen 23000, Vietnam*

Abstract. This article introduces a study on the optimal calculation of gear ratios of a two-stage helical reducer. In the study, the acreage of the cross section of the reducer was chosen as the objective function of the optimization problem. Also, the design equation for pitting resistance of a gear set was investigated. Furthermore, the equations on moment equilibrium condition of a two-stage helical reducer and their regular resistance condition were analysed. A computer program was performed to determine the optimal partial transmission ratios for getting the minimal acreage of the cross section of the system. From the results of the computer program, two models for determining the optimal gear ratios of a two-stage reducer were proposed. Using these models, the gear ratios can be calculated accurately in a simple way.

1 Introduction

In optimum reducer design, the calculation of the gear ratios of the reducer plays a very important role. The reason is that the gear ratios strongly affect the size, the dimension, the mass, and the cost of the reducer. Subsequently, optimal determination of the gear ratios of reducers has been chosen by various researchers in their studies.

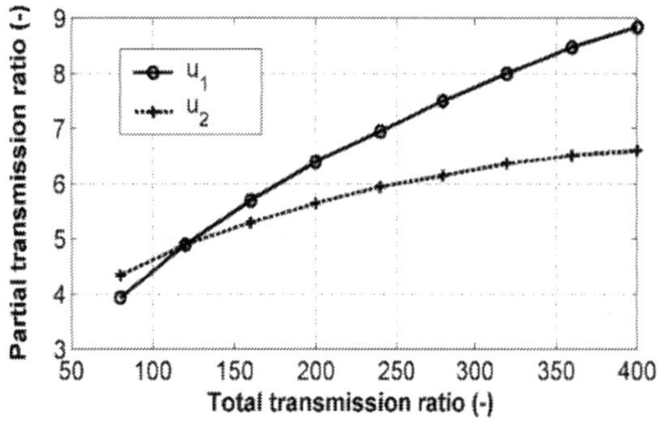

Figure 1. Gear ratios of step 1 and 2 versus the total gear ratio [1]

Up to now, there have been a number of optimization studies on determination of gear ratios of helical reducers. The gear ratios have been determined for two-stage reducers [1, 2, 3 and 4], three-stage reducers [4, 5, 6, and 8], and four-stage reducers [4, 8, 9 and 10]. Also, the gear ratios have been found using different methods including designed experience method [11], graph method [1, 2 and 4], "practical method" [2] and model method [5, 6, 7, 8, 9 and 10].

In designed experience method, the gear ratios can be found based on experiences in design and the gear ratios are not optimum. For example, for two-stage helical reducer, the relation between gear ratios of the first step u_1 and the second step u_2 follows the following equation [11]:

$$u_1 = (1.2\text{K } 1.3)\cdot u_2 \tag{1}$$

For coaxial reducers, the gear ratios can be determined by the following equation [12]:

$$u_1 = u_2 \approx \sqrt{u_t} \tag{2}$$

Where, u_1 and u_2 are gear ratios of the first and the second stage; u_t is the total ratio of the reducer.

In the graph method, the gear ratios were determined graphically. For example, for three-stage helical reducers, the gear ratios of the first stage and the second stage can be found from the graph in Figure 1 [1]. For coaxial reducers, the gear ratio of the first stage u_1 can be determined from the total ratio u_h based on graph in Figure 2 [1]. The graph method was also used by Trinh

Chat [13] for finding optimum gear ratios for two, three and four-stage helical reducers.

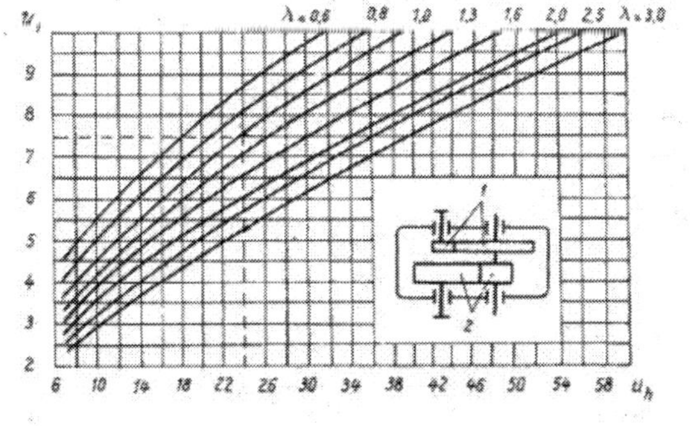

Figure 2. Gear ratio of step 1 versus the total gear ratio [1]

G. Milou et al. [2] proposed "practical method" for determining optimum gear ratios. In the "practical method", the gear ratios were found from practical data. For instance, based on the data from reducer factories, it was found that the weight of a two-stage reducer is minimal when the center distance ratio of the second stage to that of the first stage was from 1.4 to 1.6 [2]. After that, the optimal gear ratios were introduced in the tabulated form. For two-stage bevel helical gearboxes, the ratio between the bevel step and cylindrical step was 1.12 to 1.4 [2].

In the model method, the optimum gear ratios are calculated by models for different objectives. There have been quite a lot of studies which follow this method. In [5] and [14], the gear ratios were found in order to get the minimal cross section dimension of the gearbox. In [6], [9], [10] and [15], the objective for optimization problem was the minimum mass of gears. In [7], [16] and [17], the minimum gearbox length was chosen as the objective of the optimization problem for determining optimum gear ratios.

From above analysis, it is clear that till now there have been many studies on the determination of gear ratios of helical reducers. However, there is still lack of research on two-stage helical reducer which has the objective of minimal acreage of cross-section of the reducer. This paper presents a study on the optimal determination of the gear ratios for a two-stage helical reducer in which the objective is the minimal acreage of cross-section.

2 Optimization problem for determining optimal partial transmission ratios

The objective of the optimization problem is to determine the optimal gear ratios in order to get the minimal acreage of cross section of a two-stage helical reducer. From Figure 3, the acreage of cross section of the reducer A can determination on two dimensions L and h and it can be calculated as:

$$A = L \cdot h \tag{3}$$

In which, L and h are determined as (see Figure 2):

$$L = \frac{d_{w11}}{2} + a_{w1} + a_{w2} + \frac{d_{w22}}{2} \tag{4}$$

$$h = \max\left(d_{w21}, d_{w22}\right) \tag{5}$$

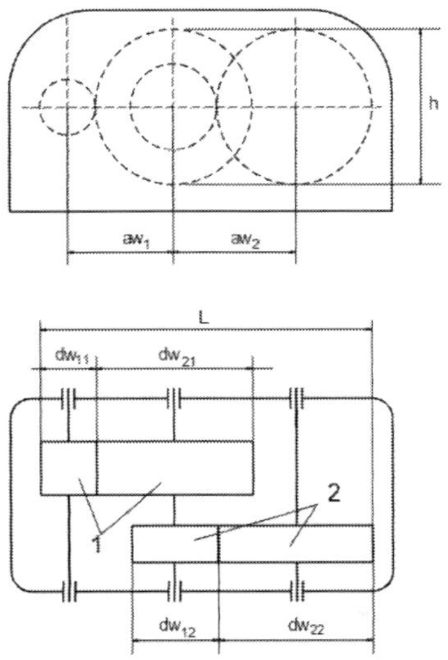

Figure 3. Calculation schema for optimum determination of gear ratio

In the above equations, a_{w1} and a_{w2} are the center distances of the first and the second stage (mm); d_{w21} and d_{w22} are the gear pitch diameters of the first and the second stage (mm); d_{w11} is the pinion pitch diameter of the first step (mm);

For a two-stage reducer, the diameters d_{w11}, d_{w21} and d_{w22} can be determined as the following equations [18]:

$$d_{w11} = \frac{2 \cdot a_{w1}}{u_1 + 1} \tag{6}$$

$$d_{w21} = \frac{2 \cdot a_{w1} \cdot u_1}{u_1 + 1} \tag{7}$$

$$d_{w22} = \frac{2 \cdot a_{w2} \cdot u_2}{u_2 + 1} \tag{8}$$

Also, for a two-stage reducer we have:

$$u_2 = \frac{u_t}{u_1} \tag{9}$$

In equations (6), (7), (8) and (9), u_1 is the gear ratio of the first stage; u_2 is the gear ratio of the second stage; u_t is the total reducer ratio.

From (4), (5), (6), (7), (8) and (9), dimensions L and h can be expressed as functions of center distances and gear ratios of two stages. Therefore, the following equations are given:

$$L = f\left(a_{w1}, a_{w2}, u_1, u_t\right) \tag{10}$$

$$h = f\left(a_{w1}, a_{w2}, u_1, u_t\right) \tag{11}$$

Thus, the optimization problem is defined as:

$$\text{minimize } A = L \cdot h \tag{12}$$

With the following constraints

$$5 \le u_t \le 40 \tag{13}$$

From (10), (11), (12) and (13), it is clear that for solving the optimization problem it is necessary to calculate the center distances of the first stage a_{w1} and the second stage a_{w2}.

2.1 Determining the center distance of the first step

From the contact strength condition, the center distance of the first step can be calculated by the following equation [18]:

$$a_{w1} = K_a \cdot \left(u_1 + 1\right) \cdot \sqrt[3]{\frac{T_{11} \cdot k_{H\beta}}{\left[\sigma_H\right]^2 \cdot u_1 \cdot \psi_{ba1}}} \tag{14}$$

Where,

- $K_{H\beta}$ is the contact load ratio for pitting resistance; for the first step of a two-stage helical reducer $k_{H\beta} = 1.02 \div 1.28$ [18]. Therefore, we can choose $k_{H\beta} = 1.1$;
- $\left[\sigma_H\right]$ - the allowable contact stress (MPa); In practice, $\left[\sigma_H\right] = 350\text{K } 410$ (MPa) and we can choose $\left[\sigma_H\right] = 380$ (MPa);
- k_a is the material coefficient; As the gear material is steel, k_a=43 [18];
- ψ_{ba} is the coefficient of wheel face width; for the first step of a two-stage helical reducer $\psi_{ba} = 0.3$;
- T_{11} is the turning moment on the pinion of the first stage ($N.mm$);

Substituting the above values into equation (14), the value of the center distance of the first stage can be found:

$$a_{w1} \approx 1.2639 \cdot \left(u_1 + 1\right) \cdot \sqrt[3]{\frac{T_{11}}{u_1}} \tag{15}$$

From the moment equilibrium condition of a two-stage helical reducer and their regular resistance condition, the output turning moment T_{out} is can be calculated by the following equation:

$$T_{out} = T_{11} \cdot \eta_{hg}^2 \cdot \eta_{be}^3 \cdot u_t$$

Wherein, η_{hg} is the helical gear transmission efficiency (η_{hg} is from 0.96 to 0.98 [18]); η_{be} is the transmission efficiency of a pair of rolling bearing (η_{be} is from 0.99 to 0.995 [18]). Choosing $\eta_{hg} = 0.97$ and $\eta_{be} = 0.992$ [18] gives

$$T_{11} = 1.0887 \cdot \frac{T_{out}}{u_t} \tag{16}$$

2.2 Determining the center distance of the second step a_{w2}

Calculating as in section 2.1 and with the note that the coefficient of wheel face width for the second stage $\psi_{ba} = 0.35$, the center distance of the second stage is calculated by:

$$a_{w2} \approx 1.2006 \cdot \left(u_2 + 1\right) \cdot \sqrt[3]{\frac{T_{12}}{u_2}} \tag{17}$$

Also, for the second step we have

$$T_{out} = T_{12} \cdot \eta_{hg} \cdot \eta_{be}^2 \cdot u_2 \tag{18}$$

Choosing $\eta_{hg} = 0.97$ and $\eta_{be} = 0.992$ as in section 2.1 gives

$$T_{12} = 1.0476 \cdot \frac{T_{out}}{u_2} \tag{19}$$

3 Results and discussions

Based on Equations (12) and (13), a computer program was performed to determine the optimal partial transmission ratios for getting the minimal acreage of the cross section of the system. The following data were used in the optimization program: $u_t = 5\text{K } 40$; $T_{out} = 10^6\text{K } 10^7$.

From the results of the optimization program, the relation between the gear ratios of two stages (u_1 and u_2) with the total reducer ratio u_t is described as in Figure 4. From the figure, it is clear that with the increase of the total ratio the gear ratio of the first stage grows faster than that of the second stage. This is because the torque on the pinion of the first stage is lower than that of the second stage. It was also found that the output torque T_{out} does not affect the gear ratios.

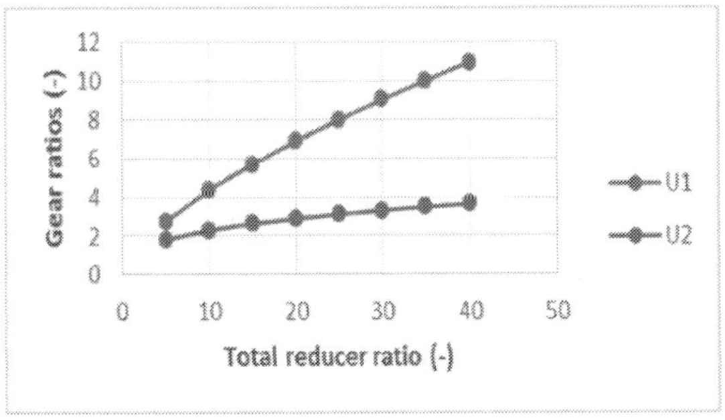

Figure 4. Gear ratios versus total reducer ratio

From the results of the program, the following regression model were proposed in order to determine the gear ratio of the first stage:

$$u_1 = 0.9365 \cdot u_t^{0.667} \qquad (20)$$

The above equation fit very well with the data. The coefficient of determination for the equation was $R^2 = 1$.

Equation (20) is used to calculate the gear ratio of the first stage of the reducer. After determining u_1 , the gear ratio of the second stage is determined by:

$$u_2 = \frac{u_t}{u_1} \qquad (21)$$

4 Conclusions

The minimal cross-sectional acreage of a two-stage helical reducer can be found by optimal determination of gear ratios of two stages of the reducer.

The models for calculation of the gear ratios of the helical reducer for getting the minimal cross-sectional acreage of the reducer were proposed.

The gear ratios of the reducer can be calculated accurately and simply by using explicit equations.

Acknowledgements

The work described in this paper was supported by Thai Nguyen University of Technology for the project N0 T2018-B17.

References

1. V.N. Kudreavtev; I.A. Gierzaves; E.G. Glukharev, Design and calculus of gearboxes (in Russian), Mashinostroenie Publishing, Sankt Petersburg, 1971.
2. G. Milou; G. Dobre; F. Visa; H. Vitila, Optimal Design of Two Step Gear Units, regarding the Main Parameters, VDI Berichte No 1230 (1996), p. 227.
3. Vu Ngoc Pi, A method for optimal calculation of total transmission ratio of two step helical gearboxes, Proceedings of the National conference on Engineering Mechanics, Ha Noi (2001), p. 12.
4. A.N. Petrovski, B.A. Sapiro, N.K. Saphonova, About optimal problem for multi-step gearboxes (in Russian),Vestnik Mashinostroenie, No. 10 (1987), p. 13.
5. Vu Ngoc Pi, Nguyen Khac Tuan, Optimum determination of partial transmission ratios of Three-step helical gearboxes for getting minimum cross section, Journal of Environmental Science and Engineering A 5 (2016) 570-573.
6. Vu Ngoc Pi, Nguyen Dang Binh, Vu Quy Dac, Phan Quang The, "Optimal Calculation of Total Transmission Ratio of Three-Step Helical Gearboxes for Minimum Mass of Gears." (In Vietnamese) Journal of Science and Technology of 6 Engineering Universities (2006), p. 91.
7. Vu Ngoc Pi, A new study on optimal calculation of partial transmission ratio of three-step helical reducers, The 3rd IASME / WSEAS International Conference on Continuum Mechanics, Cambridge, UK (2008), pp. 10-13.
8. Romhild I. , Linke H., Gezielte Auslegung Von Zahnradgetrieben mit minimaler Masse auf der Basis neuer Berechnungsverfahren, Konstruktion 44, 1992, pp. 229- 236.
9. Vu Ngoc Pi, Optimal determination of partial transmission ratios for four-step helical gearboxes with firt and third step double gear-sets for minimal mass of gears, Applied Computing Conference (ACC' 08), Istanbul, Turkey, May 27-30, 2008.
10. Le Xuan Hung, Vu Ngoc Pi, Nguyen Van Du, Optimal calculation of partial transmission ratios of four-step helical gearboxes with second and fourth-step double gear-sets for minimal mass of gears, The international symposium on Mechanical Engineering (ISME Ho Chi Minh city, Vietnam, 21-23, September 2009.
11. Nguyen Trong Hiep, Nguyen Van Lam, Design of machine elements (in Vietnamese), Educational Republishing House, Hanoi, 1979.
12. C.A. Trernapxki, G.A. Trekharev, Design of Mechanical Transmissions (in Russian), Mashinostroenie Publish, Moskova, 1984.
13. Trinh Chat, Optimal calculation the total transmission ratio of helical gear units (in Vietnamese), Scientific Conference of Hanoi University of Technology (1996) 74-79.
14. Vu Ngoc Pi, A new study on optimal calculation of partial transmission ratios of three-step helical reducers for getting minimal cross section dimension, The 2nd WSEAS International Conference on Computer Engineering and Applications (CEA'08), Acapulco, Mexico January 25-27, 2008, pp. 290-293.
15. Vu Ngoc Pi, Optimal determination of partial transmission ratios of three-step helical gearboxes with first and third-step double gear-sets for minimal mass of gear, American Conference on Applied Mathematics (MATH'08), Harvard, Massachusetts, USA March 24-26, 2008, pp. 385-388.

16. Vu Ngoc Pi, Optimal calculation of partial transmission ratios for four-step helical gearboxes with first and third-step double gear-sets for minimal gearbox length, American Conference on Applied Mathematics (MATH'08), Harvard, Massachusetts, USA March 24-26, 2008, pp. 29-32.
17. Vu Ngoc Pi, Optimal calculation of partial ratios of helical gearboxes with second-step double gear-sets, The 3rd IASME / WSEAS International Conference on Continuum Mechanics, Cambridge, UK (2008), pp. 63-66.
18. Trinh Chat, Le Van Uyen, Design and calculus of Mechanical Transmissions (in Vietnamese), Educational Republishing House, Hanoi, 1998.

Creative Design Of Sitting Hug Machine In The Treatment Of Students With Autism

Jih-Sheng Lo [1,2], Shyh-Chour Huang [1]

[1] *Department of Mechanical Engineering, National Kaohsiung University of Applied Sciences, Kaohsiung 80778, Taiwan R.O.C.*
[2] *Department of Digital Content and Technology, National Taichung University of Education, Taichung 40359, Taiwan R.O.C.*

Abstract. In recent years, the increasing number of autistic patients has attracted the attention of governments around the world. Normerally, experts will use counseling strategies, drug treatment methods, or the Squeeze Machine, which designed on the basis of the sensory integration theory, to improve those autistic patients' stereotypical behaviors performance. However, we found the existing machine on sale in the market is too bulky, more expensive and not user-friendly for those autistic students to use. Therefore, we had used expert interviews, practical operations, product analysis and then had performed several design improvements experiments, and had successfully completed the innovative prototype named 'Sitting Hug Machine' in an earlier research, and this design has won Taiwan's invention patent certificate recently. The purpose of this paper is to introduce this new design and verify its enhancement and therapeutic effect. We adapted the model of A-B-A-B reversal of Single-Subject Experimental Design(SSED) with two junior high-school autistic students, and we got a lot of good results, including improvement of stereotypical behaviors performance, more flexible operation, lower machine cost and miniaturization of machine volume, etc. Finally, we hope that the results of this study can be further promoted to benefit more autistic patients and improve their stereotypical behaviors.

1 Introduction

According to official government statistics, the patients with autism in Taiwan had reached 13,684 by 2017 [1]. In an autism survey showed more than 80% of respondents lacked independent family living skills, that is, they needed more intensive household caretaking [2]. For this reason, we hope the research can provide an effective treatment to help autistic students improve their living quality and learning outcomes and also develope more capabilities. Hence, their family can also efficiently ease the burden of caretaking. Ayres was a pioneer in the treatment of autistic children and the first to propose a theory of sensory integration. Then she further developed an intervention that provides autistic students with a deep pressure on their skin to give them a calming effect [3-6]. Grandin (1992) developed a Squeeze Machine, trying to put the above theory in practice, with its remedy effects described by Ayres validated through a series of experimentations [7-8]. We examined the Squeeze Machine manufactured by Therafin Corporation (as shown in Fig. 1) that was based on Grandin's concept and found that the machine itself has some designs which are not good enough and the some operations are also not patient-friendly. This paper is to introduce this prototype, which we had finished and was named as Sitting Hug Machine as shown in Fig. 2 [9] in an earlier research, and then to verify the therapeutic effectiveness for autistic student regarding this design.

Figure 1. The Squeeze Machine of Therafin Corporation [10]

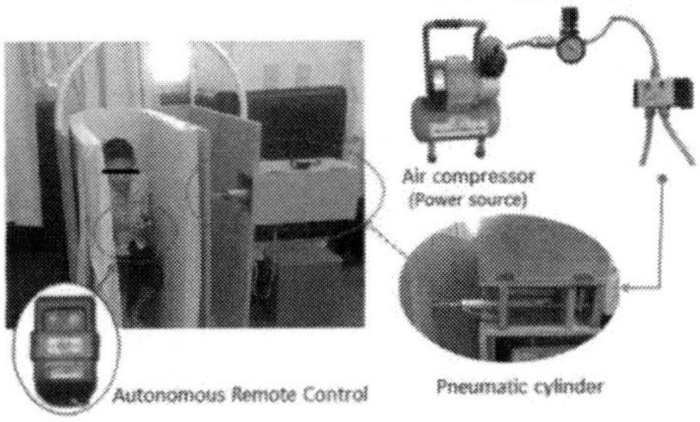

Figure 2. Prototype of the Sitting Hug Machin

2 Developing The Sitting Hug Machine

In an earlier study of this research, we first examined the Squeeze Machine of Therafin Corporation, and get a conclusion that it is too bulky, expensive and not patient-friendly, etc. Then a series of design and development work were carried out and then an improved design was called 'Sitting Hug Machine' [9]. Ayres (1979) published a theory of Sensory Integration Therapy, and through a series of experimental verification proved it is more effective than the traditional way: through counseling strategies and drug treatment etc. [8]. Then on the basis of this theory, Grandin (1992) designed the first Squeeze Machine, which was the mass produced by Therafin Corporation. This research has selected this machine as a reference model for the innovative design from the beginning [9]. Therefore, the theoretical basis of the Sitting Hug Machine design is based on the Sensory Integration Theory [6]. As Kanner (1943) found, autistic people have three common characteristics: (1) over-attachment to an object and failure in social interaction; (2) late development or inability in language using; and (3) limited range of repetitive and stereotyped behaviors [11]. Thus American Psychiatric Association (2000) defined these three characters as the core diagnostic features of autism [12]. Stereotypes are the only behavioral characteristics of the three diagnostic features of autism and are therefore the most curable in childhood of autistic patients [13]. Therapists adopt strategies that can effectively improve the stereotypical behavior of students with autism [14]. This research will use the extent of stereotypical behavior as measure indicators of the effectiveness of treatment for autism. Anna Jean Ayres (1968) first found that "learning disabilities" or "Dysfunction in Sensory Integrative Processes" is the root cause of autism, which she later named it "Sensory Integrative Dysfunction" [5], and proposed the Sensory Integration Theory [6]. By applying firm and deep pressure on autism patients' skin, it can ease their tactile defensiveness while increasing their discriminative responses significantly [7]. Therefore, Granting (1992) developed an Squeeze Machine as a remediation tool that provides a deep-pressured hug for autistic students during therapies.

3 Design Sitting Hug Machine

Earlier Studies in this Study, We first examined Therafin's Squeeze Machine and conducted in-depth interviews with therapists and patients of the machine. Our conclusion is that it has the following shortcomings: (1). The operating posture is not comfortable, which the patient must squat or tummy, and Children who are too small will not be able to apply; (2). The machine is too bulky and heavy, which needs to be placed in a large space and cannot easily move; (3). Equipment is imported Goods, and the price is very expensive;(4). Non-autonomous control methods, which is a action control box, design in the forefront of the machine outside; and (5). Uneasily adjust the squeeze point with the responding part of body. Thus, an innovative design of the 'Sitting Hug Machine' was finished as an improvement to the above shortcomings of this 'Squeeze Machine' [9]. The prototype machine (as shown in Figure 3) uses a thickness of 24mm poly-wood frame (which can also be made of lightweight metal truss frame).

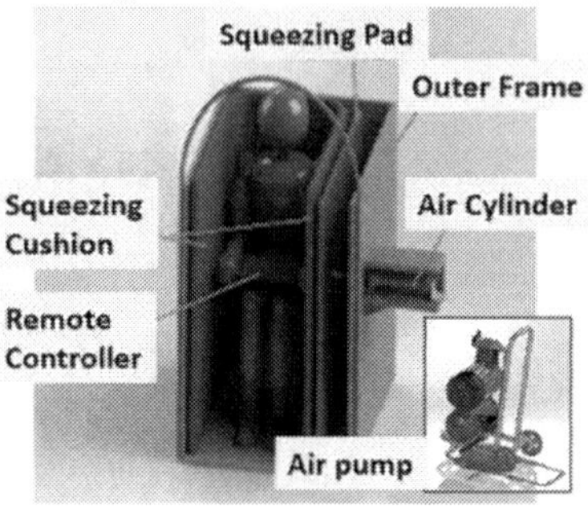

Figure 3. CAD modelling of the Sitting Hug Machine and air pump

The front door is framed by a round stainless steel tube to enhance rigidity of the whole structure and has a pneumatic pump powers to push/pull the press plate. In addition, in normal situation, the patient can sit comfortable on the seat and completely control the squeeze intensity and duration through the wireless remote control, so that the patient can feel much better and relieved. After the patient operating, the cylinder (E) pushes the squeezing pad (F-F) to swing around the center axis (C) to apply the pressing pressure on the pressing pad to give the patient a feeling of hug-like deep pressure on the both sides of shoulders or hip joints of buttocks. In addition, the swinging angle of the pressing pad to the patient can be adjusted according to the patient's feeling. By changing the position of the two squeeze regulators (A and B), the pressure exerted on the shoulders and buttocks can be adjusted separately. In addition, the patient may also take different postures, such as sitting down, standing, squatting or other gestures, according to the squeezing position, and then adjust squeezing the whole body, half-length or single hands or feet to achieve a deep feeling of comfort. Patients can also control the squeeze by the remote control the length of time or rhythm as shown in Figure 4.

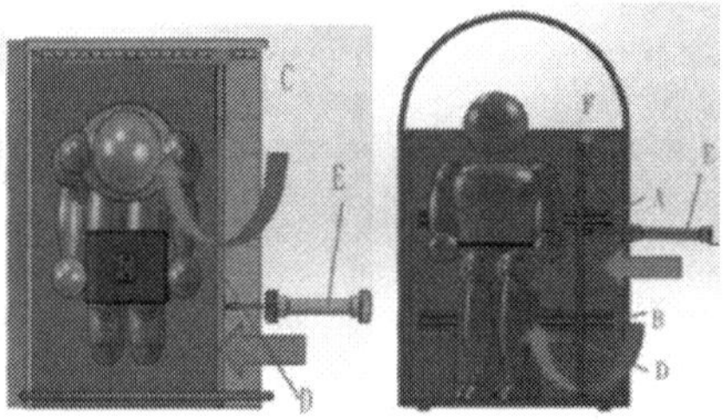

Figure 4. CAD modelling of the Sitting Hug Machine

In summary, the innovation of this innovative design hug machine has the following points: (1).more patient feeling comfortable： Wang (2005) reported her research result, indicating that sitting posture for the patient could be a better design of this machine. This is also the reason of the naming as 'Sitting Hug Machine'; (2).more compact size: the original machine design made the patient sit rather than tummy, so the machine was almost half the size of the 'Squeeze Machine'.; (3).Cheaper cost: According to the web page of Therafin Corporation, the price of Squeeze Machine is 4,525 USD, while the cost of our Sitting Hug Machine prototype is about 600 USD approximately;(4). Autonomous control: Autonomous control: While using the Squeeze Machine, the controller is a therapist. If the patient has a hand-held remote controller, and can completely autonomous control the machine, then he or she will feel safer and more comfortable; (5). Easier to adjust: As shown in Figure 3, the squeezing pad (F-F) on the patient's swing angle can be adjusted depending on the patient's feeling. By changing the position of the two squeeze regulators (A and B), the pressure exerted on the shoulders and buttocks of the patient can be adjusted respectively, which gives the patient an extraordinary comfortable feeling. Besides, the patient's posture of squeezing, regardless of sitting down, standing, squatting etc., if it is safe, there will be no limitations

4 Design validation

Although the results of our earlier study on the Sitting Hug Machine demonstrated its enhanced effectiveness through the initial in-depth interviews, its validity remains to be further verified in our prototype patient tests. The main purpose of this article is to complete this part of the study:

4.1 Methodology of experimentations

In this study, the model of A-B-A-B reversion experimental design model of SSED was adopted to verify the design improvement of the Sitting Hug Machine for Ergonomic and Remedial Ratings. While the Remedial Ratings measured by observing the reduction of stereotypic behavior of the subjects, the Ergonomic Ratings were also measured simultaneously by the questionnaire. Two junior high school students who were rated as having severe and moderate autism respectively enlisted as the subjects in these tests. The test data was further interpreted through the visual analysis and C statistics to investigate the effectiveness of using the Sitting Hug Machine as the therapeutic tool to reduce stereotypic behaviors of the subjects. The structure of the research method for this experiment was shown in Fig. 4.

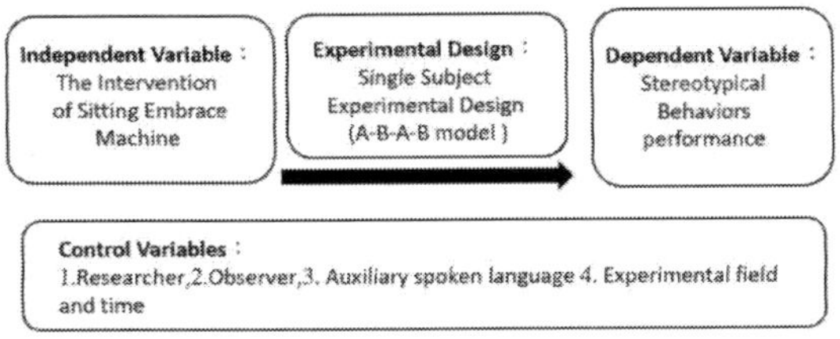

Figure 4. Structure of the research method in this experiment

4.2 Experimental results

Due to the limited source of autism cases, it is hard to obtain a large number of experimental samples. Therefore, we took the Purposive Sampling method to investigate the parents of 7 students who had used a Sitting Hug machine and 11 special education teachers who served in the same school. A total of 18 respondents, including 5 males and 13 females, obtained the data shown in Table 1. The test results of the above experiments in terms of Ergonomic and Remedial ratings revealed the facts as follows:

(1). Ergonomic rating: The "Usability" rating of this improved design is 4.10 and 4.03 for its 'Safety', respectively. These results show that the experimental participants were satisfied with the practicality of this machine. However, the score of " accessibility " was 3.92, indicating the need to further strengthen this work.

Table 1. Mean statistical score of operational perception survey.

	Gender		Title		average value
	Male	**Female**	**Teacher**	**Parents**	
Usability	4.36	3.92	3.98	4.14	4.10
accessibility	4.09	3.79	3.82	3.96	3.92
Safety	4.13	3.96	3.98	4.05	4.03

(2). Intervention effectiveness analysis : In this study, two subjects were recruited at each stage (baseline period (A1), first interventional period (B1), reversion stage (A2). The data of the second intervention period (B2) are presented with the correct rate curve (Fig. 5, 6); and the visual analysis method is further adopted. The testers were included in the summary table during each stage and between stages; they were tested with C-statistics to investigate the effect of Sitting Hug Machine intervention on stereotypical behaviors of the two subjects.

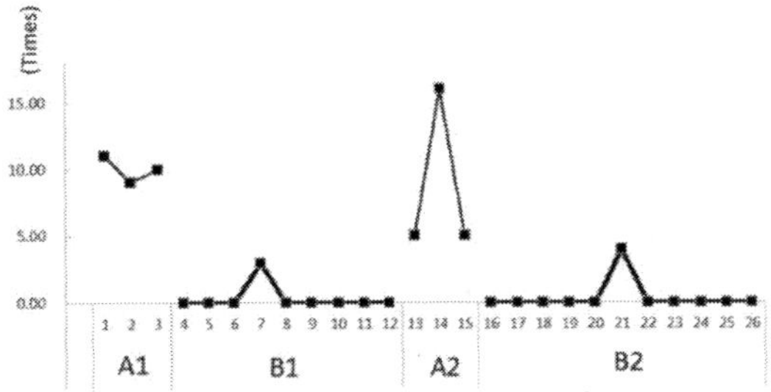

Figure 5 curves of the number of stereotypical behaviors before and after using a Sitting Hug Machine (subject A)

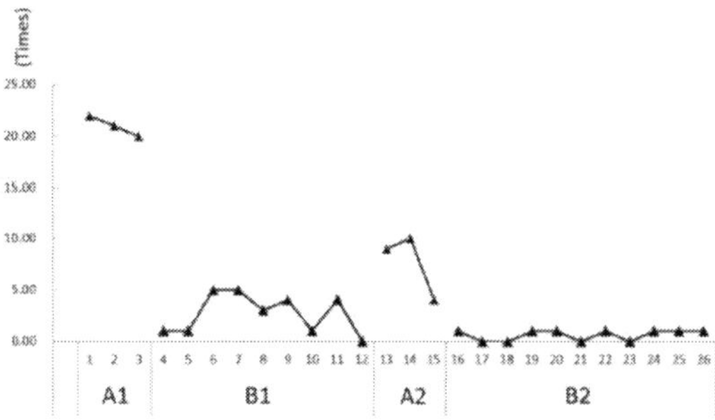

Figure 6 curves of the number of stereotypical behaviors before and after using a Sitting Hug Machine (student B)

Regarding the remedial effect of reducing stereotyped behavior in subjects, subject A reduced by 9.67 times and subject B by 18.33 times. Therefore, the z-scores of C statistic in the first stage of experimental design A1 + B1 are 2.73 and 2.78, respectively, which are both greater than 0.01, reaching a significant level. This result indicated that the remedial effect of the Sitting Hug Machine was confirmed. Then in the second stage, the reductions in stereotypic behaviors are 7.22 and 7.11 times, and the z-scores of C statistic in the second stage of reversal and experimental designs A2+B2 are 1.12 and 3.06 respectively, both being more than 0.01 again and thus reached the level of significance. This result further confirms the remedial effect of the Sitting Hug Machine.

4.3 Discussion

In summary, which the test results described above, the following points on this research could be confirmed:

(1). Compared with the Squeeze Machine, the improvements of the Sitting Hug Machine regarding the Experimental participants feeling more comfortable, the compact size and the light weight, the cheaper cost, and the autonomous control were all confirmed by therapists and patients. Only the enhancement in easy adjustment requires further advancements. Therefore, the design of Sitting Hug Machine could be assessed as successful.

(2). The remedial effectiveness of the Sitting Hug Machine is validated. The results are in line with the findings of Edelson et al. (1999) that is beneficial to increase the depth of pressure by the calming effect on children with autism.

(3). According to the tutor's observation records, the degree of concentration of subjects was significantly improved. However, the supervisor's observation records will be more subjective. Therefore, in order to further verify this result, more objective data may be needed.

5 Conclusions

This paper had successfully validated that the enhanced design of the Sitting Hug Machine developed in our former study is effective. We had made quite a dramatic improvement on this Machine, and also highly affirmed by the students, parents and special education teachers who participated in the experiment. For they found that after the experiment, students' stereotypes behaviors have been greatly improved. We not only had exhibited effectiveness in its improvements on the existing Squeeze Machine, but also had been in its remedial effects on autistic students, and this design has won Taiwan's invention patent certificate in 2015.Therefore, this Sitting Hug Machine is recommended for students with autism, and we might further commercialize it to benefit more patients.

References

1. Ministry of Health & Welfare of R.O.C., The Disabled Population. Retrieved April 18, 2017, from http://www.fact.org.tw/index.php/politic-2/2016-10-05-08-16-50/303-8787jhhhgg67676 (2017)
2. Chang, C. F., Exploring the Function of Problematic Behaviors of Children with Autism],Bulletin of Special Education], 18, 127-150 (2000)
3. Ayres, A. J., Sensory integrative processes and neuropsychological learning disability. Learning Disorders, 3, 41–58 (1968)
4. Ayres, A. J.,Types of sensory integrative dysfunction among disabled learners. American Journal of Occupational Therapy, 26(1), 13–18 (1972)
5. Ayres, A. J., Sensory integration and the child (1st Ed.). Los Angeles, CA: Western Psychological Services (1979)
6. Ayres, A. J., Sensory integration and the child: understanding hidden sensory challenges (25th Anniversary Ed., Revised and updated by Pediatric Therapy Network). Los Angeles, CA: Western Psychological Services (2005)
7. Grandin, T., Calming Effects of Deep Touch Pressure in Patients with Autistic Disorder, College Students, and Animals. Journal of Child and Adolescent Psychopharmacology, 2(1), 63-72 (1992)
8. Edelson, S. M., Edelson, M. G., Kerr, D. C., & Grandin, T., Behavioral and physiological effects of deep pressure on children with autism: a pilot study evaluating the efficacy of Grandin's Hug Machine. The American Journal of Occupational Therapy, 53(2), 145–152 (1999)
9. Lo, J. S., Lee, K. C. K., & Wang J. X., Designing a Sitting Hug Machine for Pupils with Autism. Proceedings from KEER2016: International Conference on Kansei Engineering and Emotion Research, University of Leeds, Leeds, UK: European Kansei Group. (Listed as 048.pdf in: https://drive.google.com/drive/folders/0B7l4UtIzyAG4U3Ztc1Z0MXNkNE0) (2016)
10. Therafin Corporation, Squeeze Machine [Introduction]. Retrieved from http://www.therafin.com/squeezemachine.htm (2017, June 16)

11. American Psychiatric Association, Diagnostic and statistical manual of mental disorders (DSM-IV-TR). Washington DC: American Psychiatric Association (2000)
12. LaGrow, S. J., & Repp, A.C., Stereotypic responding: A review of intervention research. American Journal of Mental Deficiency, 88(6), 595-609 (1984)
13. Anderson, C., Behavioral Therapies: Key Interventions in ASD. Interactive Autism Network. Retrieved June 16, 2017, from https://iancommunity.org/cs/simons_simplex_community/behavioral_therapies (2012)
14. Wang, C. L., The Effects of Deep Pressure Touch Treatment of Anxiety of Children with Autism]. Master Thesis, The Department of Physical Therapy and Assistive Technology, National Yang-Ming University, Taiwan] (2005)

Investigating Parameters That Effect to Wear of Plasma Nozzle

Paphakorn Pitayachaval[1] and Muhammatsoifu Sato[2]

[1]*School of Industrial Engineering, Institute of Engineering, Suranaree University of Technology, Nakhon Ratchasima 30000, Thailand*
[2]*School of Industrial Engineering, Institute of Engineering, Suranaree University of Technology, Nakhon Ratchasima 30000, Thailand*

Abstract. Plasma cutting machine normally applies to cut metal in electrical conductivity industrial according to an accuracy dimension and rapidly time consuming. A quality of this process is depend upon a cutting surface and kerf by controlling diameter of nozzle and size of electrode. Since kerf shape is crated base on nozzle diameter, while electrode is served plasma arc. This paper presents an investigating three cutting parameters: cutting speed, pressure gas, current ampere that affect to wear of plasma nozzle. The fixed variables are a plasma-cutting machine, Hypertherm powermax 45 xp, Bindee control CNC machine and specimens (100x100 mm.). The cutting speed was holed at 200, 300, 400 mm/min. The gas pressure was controlled at 6, 7, 8 bar. The current Ampere was handled at 40, 42, 45 A. The diameters of nozzle was measured using digital microscope. The experiments conducted based on ANOVA to establish the relationship of those parameters. The nozzle wear depended upon the current Ampere, the high gas pressure while the cutting speed was not effect to nozzle wear.

1 Introduction

A metal cutting process is wildly used in general industry to cut several metal spare parts and framework such as railway, aircraft, automobile, etc. Even through there are a variety of metal cutting machines such as laser cutting, oxy-fuel gas cutting, water-jet cutting, these machine provide a cutting edge quality that contradicts with product cost. The laser cutting gives a high quality while demand more expend. However, a plasma-cutting machine gives acceptable cutting quality and production time while the product cost is admitted [1].

The plasma cutting machine is a machine to cut metal with high temperature [2]. This machine contains DC power source, pressure plume and cutting head, which composes of torch (cathode), electrode and nozzle, as shown in figure 1. The plasma art is conducted by heated gas based on extremely high temperature [3], then it transfer via nozzle to work pieces that becomes electrically conductive. After that, the metal is to be melted and removed along cutting pattern. An advantage of plasma cutting is a clean process [4] that produces dross free with minimal heat input. Plasma arc frame is conducted by nitrogen gas and oxygen gas. This frame can be used to cut a conductive material including ferrous and nonferrous metal. Since there are several parameters of acceptable quality for plasma cutting such as cutting speed, current ampere, height of cutting head and gas pressure [5]. To control process parameters that are affects quality of cutting; there are only three key parameters; cutting speed, current ampere and gas pressure, respective [6].

Beside those parameters, heat of plasma gas, Generated by the plasma beam during the cutting operation, also conducts wear of plasma device components, especially nozzle wear for the consumable part [7]. This wear introduces inconstancy frame diameter. In order to obtain the acceptable cutting quality, larger orifice nozzle and electrode were applied [8]. Moreover, variable factors that affect tool life of nozzle and electrode depended upon many factors that are a number of cutting pierces, cutting duration, material thickness [9], including degree of erosion of the nozzle around the orifice nozzle [10]. To investigate those parameters, this paper attempts to conduct experiments for cutting speed, pressure gas, and current ampere in order to identify state of parameters that affect to wear of plasma nozzle.

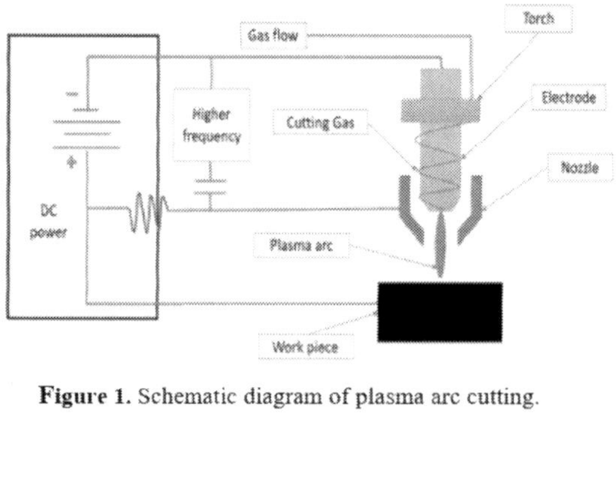

Figure 1. Schematic diagram of plasma arc cutting.

https://doi.org/10.1051/matecconf/201821301010

Table 1. The factors comparison of fuel-gas cut, laser cut and plasma cut from [11]. (Publication Section 5: Electromagnetic Heating Department of Alternative Energy Department and Efficiency http://www.dede.go.th)

Machine	Fuel-gas cut	Laser cut	Plasma cut
Quality cut	Pool cut quality get most burns and Finishing work piece	best cut quality get minas burns	Good cut quality get burns
Cut speed	Quite slowly	Fast (cut < 0.25 in)	Fast
Machine price	Around 200,000-500,000 for machine cut metal diameter 20x6.5 ft .0.4 in thickness	Around 300,000 for machine cut metal diameter 20x6.5 ft .0.4 in thickness	Around 120,000 for machine cut metal diameter 20x6.5 ft .0.4 in thickness
Consultable	Tip cutter	Protective glass, gas nozzle Dust filter	Cutting nozzle, electrode

2 Experimental

There are three factors that were investigated on this experimental: current ampere, cutting speed, gas pressure. The current ampere (40, 42, 45A), cutting speed (200, 300, 400mm/min) and gas pressure (6, 7, 8 bar) ware applied to cut metal (588GrA, 242 Type I) 10 mm. thickness. The cutting parameters were established as presented in Table 2.

Table 2. The cutting parameters for experimental

Factor	Symbol	Level			Unit
		1	2	3	
Current Ampere	A	40	42	45	A
Gas Pressure	B	6	7	8	bar
cutting Speed	C	200	300	400	mm/min

The experimental machine was the Hypertherm powermax 45 xp and the head moving control unit was applied Bindee control, as show in Fig 2. The size of specimens were 100 mm x 100 mm, as show in Fig 3. A digital microscope model Keyence VHX 700F was applied to measure diameter of nozzles after cutting, as show Fig 4.

Figure 2. Plasma cutting machine , bindee control.

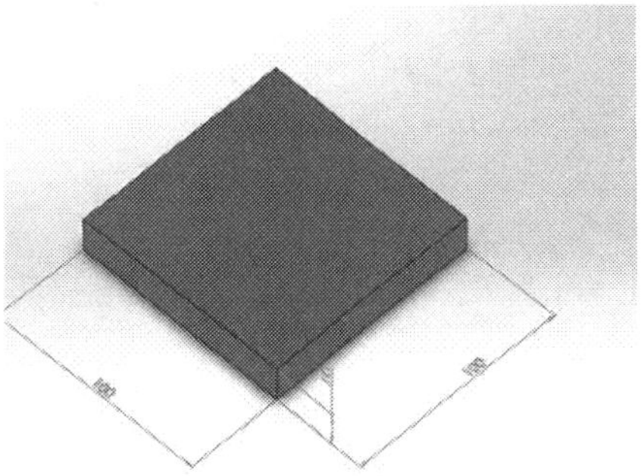

Figure 3. The size of specimen

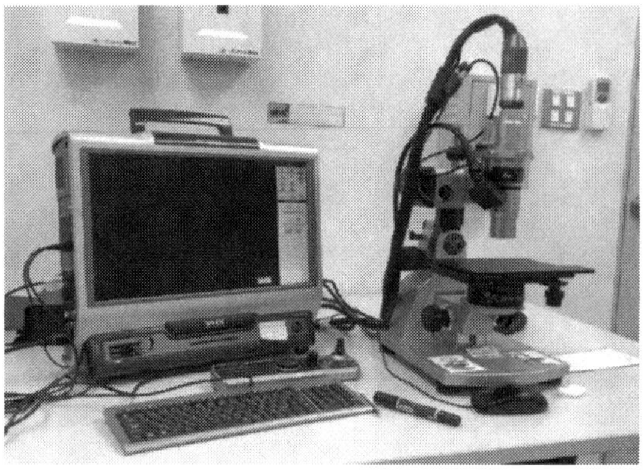

Figure 4. Digital microscope Keyence VHX 700P.

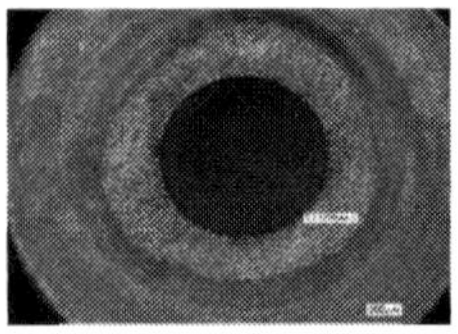

Current ampere
42 A
Gas pressure 7

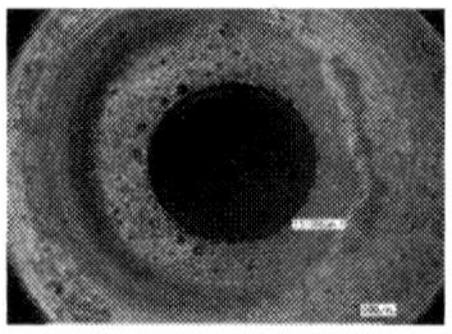

Current ampere
40 A
Gas pressure 8

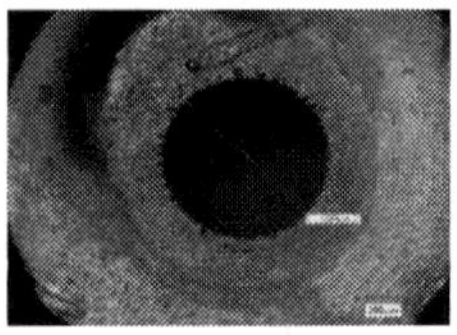

Current ampere
40 A
Gas pressure 7

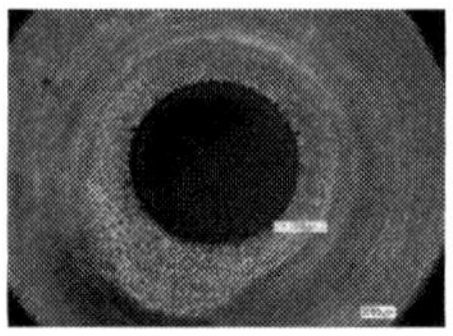

Current ampere
45 A
Gas pressure 7

Figure 5. The variation of orifice nozzle.

4 Result

Based on single parameter plasma cutting, the wear of orifice nozzle depended on current ampere and gas pressure, respectively. Beside, two factors analysis founded the interaction between current ampere and gas pressure that effect to wear orifice nozzle, as shown in Table 1. The results of the experiment was obtained from twenty-seven examples, which measure the orifice diameter to check the wear by a digital microscope, as shown in figure 5. Minitab statistic software was applied to analyze interaction effects of those parameters, as shown in table 3.

Table 3. ANOVA analysis by minitab software

Source	DF	Adj SS	Adj MS	F-Value	P-Value
Current ampere	2	0.000045	0.000023	0.62	0.562
Gas pressure	2	0.000070	0.000035	0.96	0.423
Cutting speed	2	0.000386	0.000193	5.27	0.035
Current ampere*Gas pressure	4	0.000079	0.000020	0.54	0.714
Current ampere*Cutting speed	4	0.000155	0.000039	1.06	0.437
Gas pressure*Cutting speed	4	0.000371	0.000093	2.53	0.123
Error	8	0.000293	0.000037		
Total	26	0.001399			

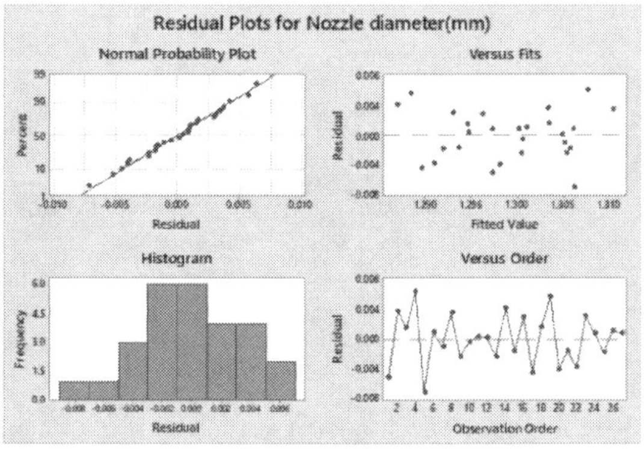

Figure 6. Residual Plots for Nozzle diameter.

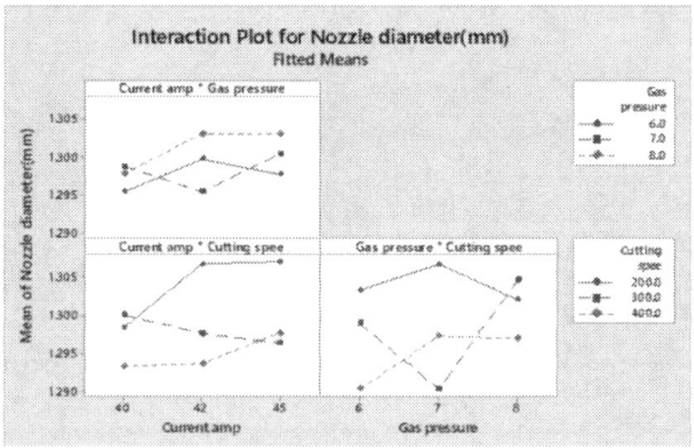

Figure 7. Interaction plot Nozzle diameter.

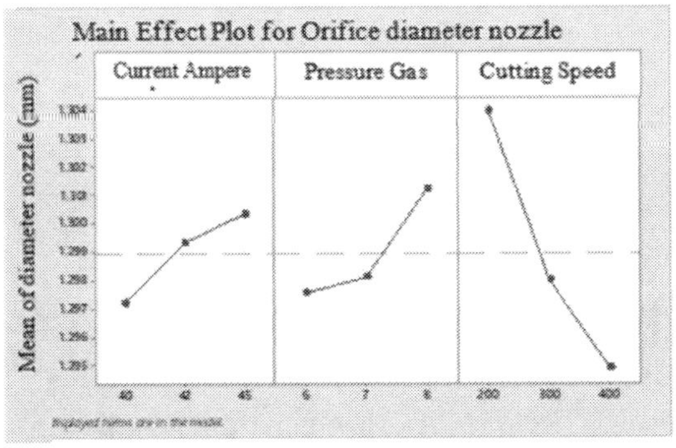

Figure 8. Main effect plots for orifice nozzle diameter.

The normal probability graphs obtained from the experiment presents the normal probability of variable, as shown in figure 6, in which the variable values are near the ideal normal line, the data from the experiments was distributed under the variation of natural causes. However, the current ampere, gas pressure and cutting speed was interaction, as shown in fig 7.

From Fig 8 The results shown that cutting speed were increased from 200 to 300 mm/min and from 300 to 400 mm/min the orifice nozzle damaged diameters were decreased rapidly. Moreover, plasma gas pressure were controlled to increase from 6 to 7 bar and from 7 to 8 bar, the orifice nozzle damaged diameter were also expanded. When current amperes were increased from 40 to 42 A and from 42 to 45 A, the orifice nozzle damaged diameter were also increased, as shown in figure 8. Therefore, the increasing of controlled parameters effect to the orifice nozzle damaged diameter.

5 Conclusions

For plasma cutting, the parameters that affects to nozzle wear were cutting speed, plasma gas pressure and current ampere. These parameters were also presented their interactions for increasing the orifice nozzle damaged diameter. However, the relationship of those parameters will be mathematical defined as the future work.

Acknowledgment

This research has been supported by the research and development supporting fund, Suranaree University of Technology.

References

1. Plasma Arc Cutting Cost S.M. Ilii1*, M. Coteaţă1** 2009
2. Parametric Study On Plasma Arc Cutting Of Al – 6063 Hardik Dholakiya1, Vikas Gondalia2 2017
3. Plasma Cutting In Atmosphere And Under Water W. Bach And A. Gruchow 1992
4. Faq: What Are The Advantages And Disadvantages Of The Principal Cutting Https://Www.Twi-Global.Com/Technical-Knowledge/Faqs/Faq-What-Are-The-Advantages-And-Disadvantages-Of-The-Principal-Cutting-Process/
5. (Experimental Investigation Of The Plasma Arc Cutting Process , K. Salonitisa,*, S. Vatousianosb 2012)
6. Analysis In Plasma Arc Cutting Of 21Cr Ferritic Stainless Steel H.Ravi Kumar And S.Thileepan 2015
7. Nozzle And Method Of Operation For A Plasma Arctorch (Richard W. Couch, Jr., Hanover; Nicholas A. Sanders, Norwich; Lifeng Luo, Lebanon; John Sobr, Enfield; Patrik Backander, Lebanon, All Of N.H. 1992)
8. A Review On Nozzle Wear In Abrasive Water Jet Machining Application H. Syazwani1, G. Mebrahitom, A. Azmir 2016
9. Optimizing Consumable Life In Mechanized Plasma Cutting online Http://Www.Thefabricator.Com/Article/Plasmacutting/Optimizing-Consumable-Life-In-Mechanized-Plasma-Cutting
10. Detecting Non-Symmetrical Nozzle Wear Naplasmaarc Torch Maciej Wlodzimierz Rogozinski.Bulleen; Subramania Ramakrishnan,Balwyn North; Ashley Grant Doolette,Wishart, All Of Australia; Nicholas A.Sanders, Norwich, Vt.; Richard W.Couch, Jr.. Hanover, N.H. 1998
11. Publication Section 5: Electromagnetic Heating Department Of Alternative Energy Department And Efficiency Http://Www.Dede.Go.Th

Design and Implementation of 3D Printing System for Continuous CFRP Composites

Ningda Han[1], Jun Cheng[1], Jiquan Yang[1], Yijian Liu[1], Wuyun Huang[1]

[1]*Jiangsu Key Laboratory of 3D Printing Equipment and Manufacturing, Nanjing Normal University, Nanjing 210042, China*

Abstract. The rapid and low-cost manufacturing of continuous Carbon Fiber Reinforced Polymer (CFRP) composites using 3D printing technology is a hot topic in the field of composite materials' research. Due to the continuity and infusibility of the long carbon fiber, a series of problems such as loosening of fiber, breakage, and nozzle clogging occurred in the printing process, which result in poor surface quality and performance in the printed product. This paper aims to solve these problems based on the researches and optimizations of three-dimensional printing technology for continuous CFRP composite components. Firstly, the coupling mechanism of continuous fiber and resin polymer in the flow path of nozzle is analyzed, the finite element simulation models of flow field and temperature field of CFRP three-dimensional printing are established by using ANSYS CFX software, and the coupling characteristics and interface performance in the printing process are studied. Then, based on the results of simulation analysis, a modification method of the surface coating film is applied, and a special modification solution is configured to modify the surface of the carbon fiber so as to increase its strength and bondability with the molten resin. Finally, the mechanical structure of the three-dimensional printing system of continuous CFRP components is designed to achieve the synchronization of printing and fiber modification. Considering the continuity of continuous carbon fiber, this paper proposed a new method of printing path design called "unicursal" for continuous CFRP parts, that is, when designing and planning a three-dimensional print path, it ensured that there is no interruption in the printing process, so as to achieve carbon fiber continuity in composite parts. The reliability and superiority of the printing system designed in this paper are confirmed by printing of the composite parts.

1 Introduction

Composite material is a new type of material synthesized from two or more materials of different properties by physical or chemical methods. Composite materials combine the advantages of different materials in terms of performance, so that the overall performance of the composite material is better than a single material to meet a variety of different needs. Continuous carbon fiber-reinforced resin-based composites are one of the most widely used materials in today's industrial fields such as aerospace and automotive manufacturing. The manufacturing process of the traditional Continuous CFRP component mainly includes three steps: First, the prepreg is prepared by a deposition method or a dipping method; then a simple-shaped composite material part is manufactured through a process of laying, pultrusion, and winding; Finally, continuous CFRP members are prepared by machining assembly, cementation and other processes[1]. The manufacturing process of this method has many disadvantages. For example, the process is complicated, secondary processing is required, and the production cycle is long; most of the production requires a mold and the cost is high; the processes such as cementing or assembly are required, and the complexity of the manufacturing process is limited.

Three-dimensional printing (also known as 3D printing, rapid prototyping, additive manufacturing, etc.) is considered to be a major breakthrough in manufacturing technology in the past 20 years, and is also known by many domestic and foreign media as a representative technology of the "third industrial revolution". Under the control of a computer, components of any complex shape can be rapidly fabricated through accurate 3D accumulation of materials based on data such as computer aided design (CAD) models or computed tomography (CT) scans of objects. Undoubtedly, 3D printing technology (Three dimension, 3D) opens up a whole new idea for the manufacture of composite materials. Compared with traditional manufacturing processes, 3D printing technology has the advantages of low cost, high material utilization, and simple process[2]. With the advent of the 3D printing process and its rapid development, it provides an effective technical approach for the low-cost rapid manufacturing of continuous CFRP components, making it possible to integrate the rapid manufacturing of CFRP components with complex structures[3]. Improving the applicability of 3D printing technology in the manufacture of composite materials has important engineering significance for promoting the cost-effective manufacture of continuous fiber-reinforced resin-based composite products.

In the 3D printing process of continuous fiber reinforced resin composites,because of the continuous carbon fiber does not melt at a high temperature and its own strength is insufficient, it often causes the problems of continuous fiber breakage and fiber blocking of the nozzle in the three-dimensional printing, resulting in the result that the surface of the product is not smooth, the precision is low, and the component cannot be formed[4-6]. Therefore, how to improve the practicality of 3D printing equipment in continuous CFRP component manufacturing has become a hot topic in the current research field.

This article mainly focuses on the above issues, researches and optimizes the three-dimensional printing technology of continuous CFRP composite components. Firstly, the coupling mechanism of continuous fiber and resin polymer in the flow path of nozzle is analyzed, and the finite element simulation model of flow field and temperature field of CFRP three-dimensional printing is established by using ANSYS CFX software. The heterogeneous coupling characteristics and interface of the composite material in the printing process are discussed. Then, based on the results of simulation analysis, a modification method of the surface coating film is applied, and a special modification solution is configured to modify the surface of the carbon fiber so as to increase its strength and bondability with the molten resin[7]. Finally, the mechanical structure of the three-dimensional printing system of continuous CFRP components is designed to achieve the synchronization of printing and fiber modification. Taking into account the continuity of continuous carbon fiber, this article proposes a new type of "unicursal" print path for continuous CFRP components, that is, when planning a three-dimensional print path, it is allowed to have no interruption in the printing process, so as to ensure the composite Continuity of Carbon Fibers in Molded Parts[8]. the three-dimensional printing of continuous CFRP components proves the effectiveness of the system designed in this paper.

2 Simulation

2.1 Model establishment

ANSYS CFX is a high-performance computational fluid dynamics (CFD) software that provides fast, robust, reliable and accurate solutions for a wide range of CFD and multiphysics applications.After meshing, it can be imported into the CFX module of the ANSYS workbench for analysis. Save it as "mesh" format and import it into CFX as shown in Figure 1(a). Then click on "setup" to start the calculation process.

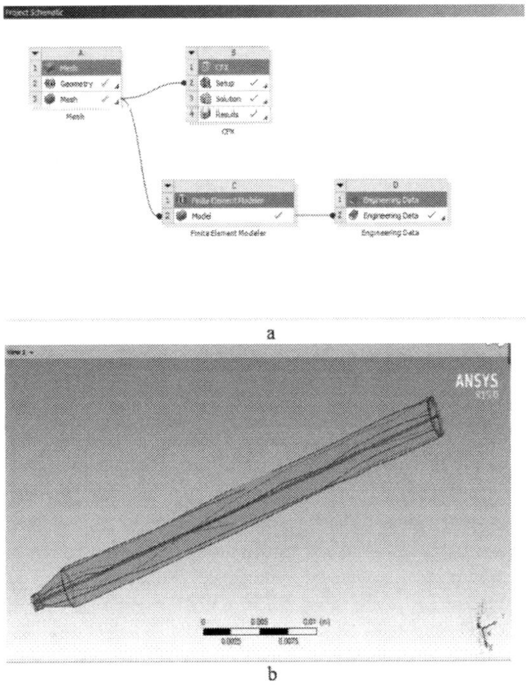

a

b

Figure 1. CFX module interface (a)and Model after importing CFX(b)

The imported model is shown in Figure.1(b). After determining the model is correct, set the boundary conditions. According to the actual situation, set the type of fluid, the material of the fluid, the material of the outer wall of the nozzle, and the inlet and outlet flow speeds.

2.2 Simulation results

Click 'solution' for iterative calculations. After the calculation is completed, enter 'result' to analyze the result.

After the calculation is completed, select post-processing. In the post-processing model, there are multiple reference indicators. However, the focus of analysis in this paper should be close to the actual production. When the fluid rotates, it takes into account the viscosity of the material and the pressure of the fluid on the carbon fiber. Effects, etc. The selected reference indexes are velocity distribution and pressure distribution.

(1)Pressure Analysis

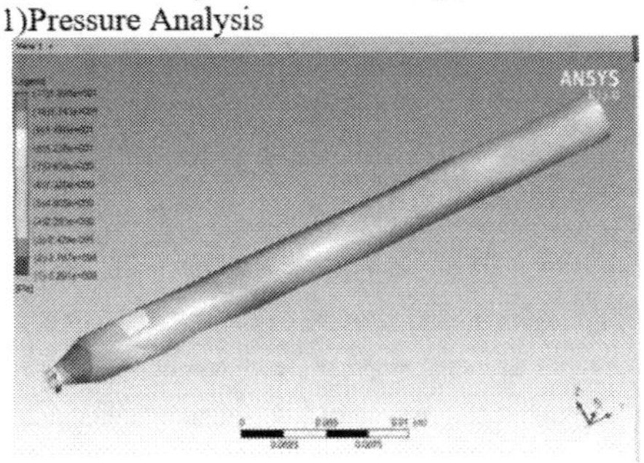

a

 https://doi.org/10.1051/matecconf/201821301011

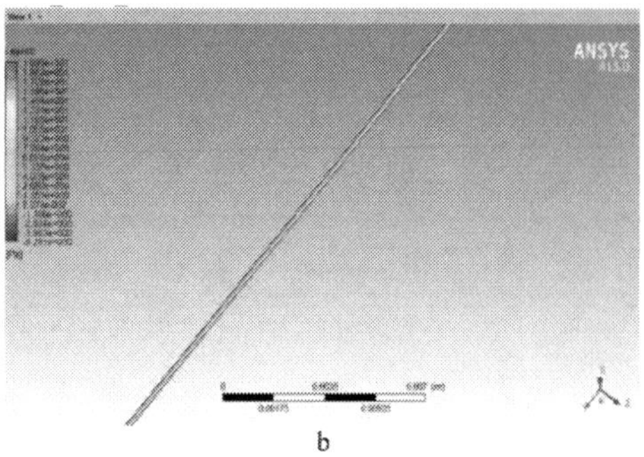

b

Figure 2. Partial pressure cloud image of nozzle(a) and Carbon fiber pressure cloud(b)

It can be seen from Figure 2(a) that the pressure on the entire wall surface is relatively uniform and less than the pressure at the nozzle, and the pressure at the nozzle presents a tendency of rising first and then decreasing, and the pressure closer to the air space is smaller. Carbon fiber is driven by the flow of molten resin to move it outward. Therefore, the pressure of the fluid on it must be considered.

Figure 2(b)shows the carbon fiber surface pressure cloud. It can be seen from the figure that the carbon fiber is surrounded by the fluid and the pressure received is uniform and small, which meets the actual needs.

(2)Speed analysis

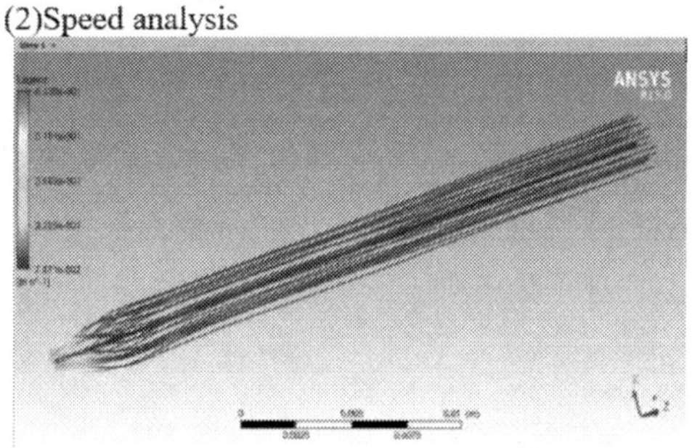

Figure 3. Velocity streamline of Fluid

Another reaction indicator of the flow field is the fluid trace. Changes in the fluid trace can reflect changes in the distribution and trends of fluid flow. That is, it can reflect the actual liquid flow in the nozzle and thus has a reference effect on the change of the flow field.

As can be seen from Fig. 4, the viscosity of the molten resin causes the flow rate of the liquid to decrease and maintains a uniform velocity in the nozzle, and the speed increases when the fluid reaches the nozzle head.In actual operation, the simulation data can be compared to determine whether the fluid inside the nozzle is blocked.

3 Process design

From the simulation results, it can be seen that in the three-dimensional printing process of the continuous CFRP component, although the force around the carbon fiber is even, the force is very small. This may be such that the carbon fiber cannot be extruded from the nozzle together with the molten resin during printing

In order to overcome the defects in the existing 3D printing technology of continuous CFRP components, this paper will design a 3D printing method for continuous CFRP components that can be simultaneously modified and printed based on the simulation results.

3.1 Process principle

The feature of this process is that during the printing process, the continuous carbon fiber is infiltrated with the modifying solution before reaching the print head, so that the surface properties of the continuous carbon fiber are improved, the continuous carbon fiber is not easily broken and can be better combined with the molten liquid resin.

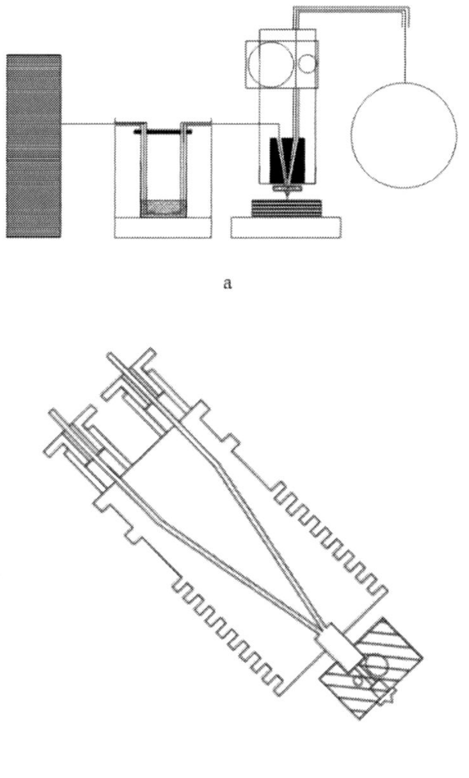

a

b

Figure. 4 Process principle (a) and print head design (b)

As shown in Figure 4(a). The process uses thermoplastic resin and continuous fiber as the printing material, and the two materials respectively enter the print head from the left and right two entrances of the print head (the internal structure of the print head is shown in Figure 4(b)).The continuous carbon fiber is infiltrated and modified by the modifying solution before entering the print head, and then the modified carbon fiber enters the heating chamber to be combined with the melted resin and is printed out. Finally, the modification of carbon fiber and three-dimensional printing are carried out simultaneously to realize the integration of composite material preparation and forming. The thermoplastic resins that can be used in this process include PLA, ABS and nylon, and the continuous fibers include carbon fiber and glass fiber.

3.2 Preparation of modified solution

The method of surface modification of carbon fiber used in the three-dimensional printing of continuous CFRP components designed in this paper is a modification method by coating a film on the surface.

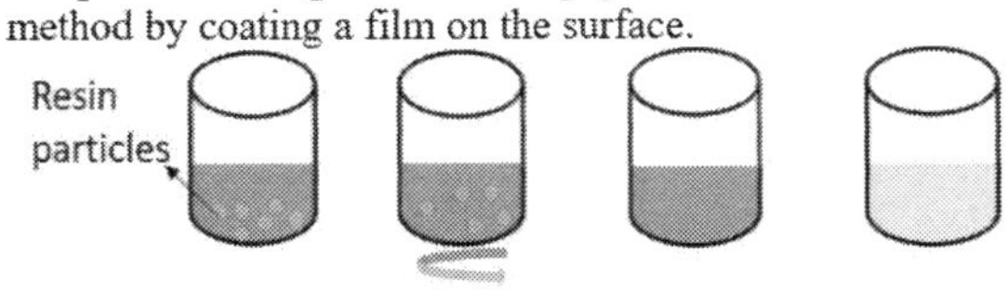

Figure.5 process of modifying the solution configuration

As shown in Figure 5, the configuration of the modified solution is divided into the following steps:

1. Pour Dichloromethane into a beaker and add resin particles (PLA or ABS) with a volume fraction of 5% to 10%. Stir with a glass rod.

2. Place the beaker on a magnetic stirrer, set the rotation speed to 3000 or more, set the temperature to room temperature, and set the stirring time to half an hour. Wait until the stirring is completed.

3. The stirred solution is filtered through a gauze to obtain a clear, viscous resin-dichloromethane solution, which is the modified solution.

After the carbon fiber is infiltrated with the modified solution, it is equivalent to the surface of the carbon fiber wrapped in a layer of resin, so the combination of the resin with the melted resin will be greatly improved, and it will be easier to move with the flow of the resin during the printing process.The methylene chloride itself is highly viscous. After the carbon fiber is infiltrated, the internal carbon fiber filaments will be more firmly bonded and will not be easily dispersed in the three-dimensional printing process. improve the quality of printed CFRO components Ultimately.

4 path planning and printing

4.1 Continuous 3D printing path planning

This paper studies the printing path planning of 3D printing process based on continuous CFRP components. Unlike traditional printing paths, the 3D printing path satisfies some new requirements due to the continuity of the carbon fibers in the continuous CFRP components. Studies have shown that the printed path not only affects the final molding quality of the composite structure, but also that the fiber orientation in the composite material has a great influence on the performance of the final molded part. Therefore, how to design the structural form of the printing member structure and the design of the internal continuous fiber to control the print path of the fiber is the key to realize the three-dimensional printing manufacturing of the high-performance continuous CFRP member. For this reason, this article has designed a " unicursal" continuous print path setting.

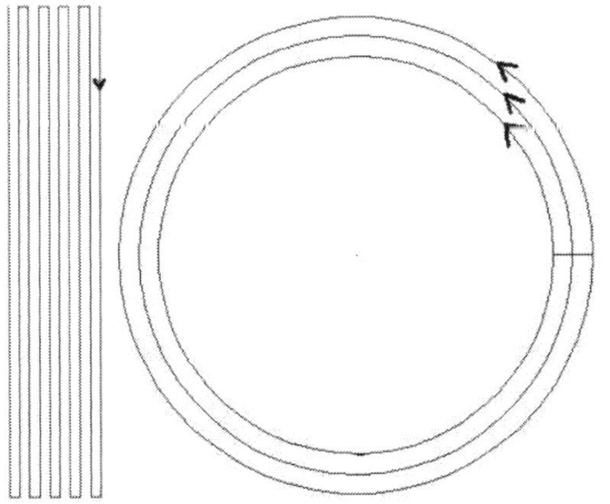

Figure. 6 Continuous 3D Print Path Planning

In the case of a three-dimensional printing of a rectangular continuous CFRP member, the print path is as shown in Figure 6. The print head moves in accordance with the uninterrupted "unicursal" path. After printing a layer, the next layer is reversed. The direction continues to print, and the final layers are stacked. The print path setting of the closed circular continuous CFRP member is as shown in Figure 6. Let the print path be printed from the inside outward or from the outside to the inside, and set a cut-in point after printing a closed loop to print the next Closed loop, the last layer of stacked forming.

4.1 Process validation

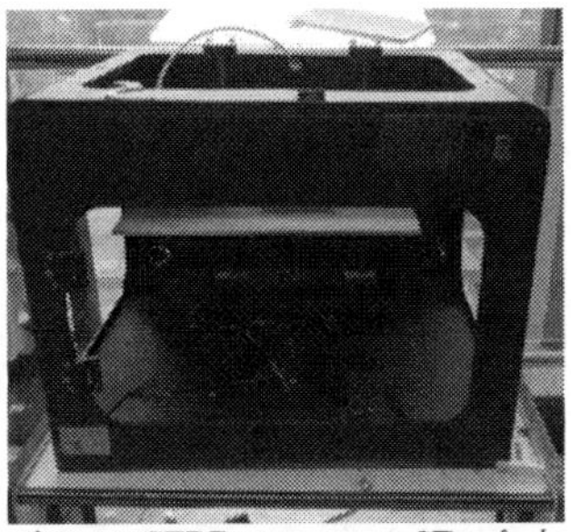

Figure 7. Continuous CFRP component 3D printing equipment

In order to verify the feasibility of the proposed process, this paper preliminarily completed the fabrication of continuous CFRP component 3D printing equipment based on the three-dimensional printing principle of FDM and the simultaneous modification and printing process of continuous CFRP components, as shown in Figure 7.

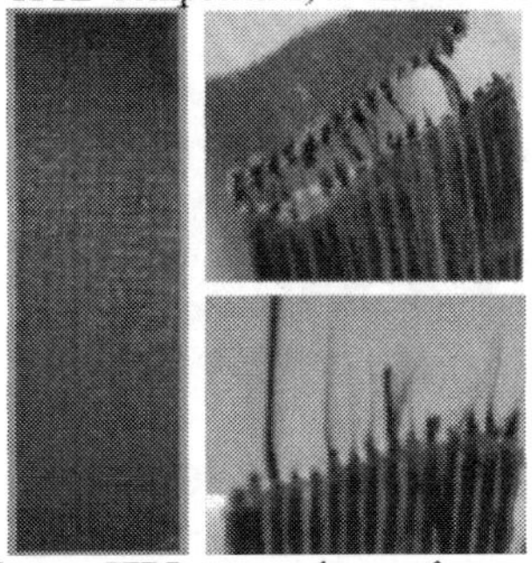

Figure 8. Continuous CFRP rectangular member

In this paper, a rectangular continuous CFRP object is printed on the device according to the "unicursal" print path, as shown in Figure 8. As can be seen from the figure, the entire component surface is fairly flat and smooth. Inside the construction, continuous carbon fibers are uniformly distributed on each layer of the member and the carbon fibers are tightly bound to the resin. Continuous carbon fibers do not be broke apart during printing.

5 Conclusion

The 3D printing system of continuous CFRP components designed in this paper can effectively improve the surface properties of continuous carbon fibers in the threedimensional printing process, improve the combination of carbon fibers and melted resin, and make the continuous carbon fiber materials well wrapped with thermoplastic resin And make continuous carbon fiber not easy to break in the printing process. Through the " unicursal" print path planning, continuous carbon fibers will not break due to discontinuities in the path. The 3D printing continuous CFRP component is prone to defects such as burrs, fiber breakage, and poor adhesion between the resin and continuous carbon fiber, which greatly improves the three-dimensional printing quality of the continuous CFRP component.

References

1. P. Mitschang, M. Blinzler, A. Wöginger, Compos. Sci. Technol, **63(14)** 2099-2110 (2003)
2. BM. Tymrak, M. Kreiger, JM. Pearce, **58(6)** 242-246 (2014)
3. F. Ning, W. Cong, J. Qiu, J. Wei, S. Compos. Part. B-Eng, **80** 369-378 (2015),
4. R. Matsuzaki, M. Ueda, M. Namiki, Sci. Rep, **6** (2016)
5. T. Semba, K. Kitagawa, US. Ishiaku, H. Hamadai, J. Appl. Polym. SCI, **101(3)** 1816-1825 (2016)
6. Q. Chen, P. Boisse, CH. Park, *Compos Struct*, **93(7)** 1692-1703 (2011)
7. X. Tian , T. Liu , C. Yang , Q. Wang, Compos. Part. A-Appl. S, **88** 198-205 (2016)
8. HL. Tekinalp, V Kunc, GM. Velez-Garcia, Compos Sci Technol, **105** 144-150 (2014)

A Review of 3D Food Printing Technology

Paphakorn Pitayachaval[1], Nattawut Sanklong[1] and Anantapoom Thongrak[1]

[1]*School of Industrial Engineering, Institute of Engineering, Suranaree University of Technology, Thailand*

Abstract. The additive manufacturing technology has been applied to directly construct physical model from 3D model without mold and die. Several industries utilize this technology to manufacture a complicated part such as automobile, aerospace including food industry. The advantage 3D food printing are ability to produce complex food model and ability to design unique pattern. A 3D food printing technique is composed of an extrusion-based printing, binder jetting and inkjet printing. The food materials such as sugar, chocolate, and cheese are used to create designed shape based on layer-by-layer. This paper presents a review of 3D food printing techniques. This review is to categorize, printability, productivity, properties of material and mechanism of 3D food printing techniques, as well as to provide the direction of future development.

1 Introduction

A three-dimensional (3D) printing, called additive manufacturing (AM), established since 1980s, have been developed and applied in variety applications for many industries. AM crates model by adding material layer by layer from a computerized 3D solid model. An advantage of AM is to construct a complexity model without mold and die, fixtures, cutting tools and coolants. The application of construction AM model have been wildly used in many fields of industry such as automotive, architecture, medical and fashion design. Including, food manufacturing also applies this technology to fabricate food design. However, a sustainable nutrition and food security are the global agenda and key themes, which are considered during, apply 3D food printing. There are several techniques to construct 3D food printing that are an extrusion-based printing, binder jetting and inkjet printing. This paper aims to review those techniques based on a printability, productivity, properties of material, effect parameters and mechanism of 3D food printing techniques. Including, the advantages and disadvantages of those techniques are also established.[1-7]

2 Category of 3D Food Printing Technique

The 3D food printing technique has been classified into three categories that are an extrusion-based printing, binder jetting and inkjet printing, as shown in figure 1.

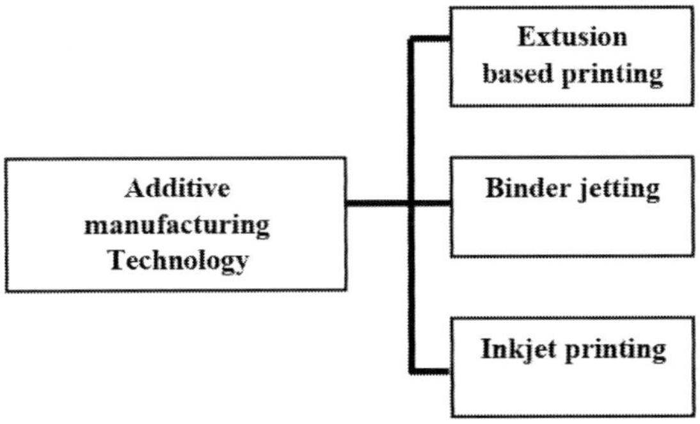

Figure 1. Classified of 3D printing technique

2.1 Extrusion-based printing

The extrusion-based printing constructs food model by extruding food through a nozzle with a constant pressure. This technique is similar to a conventional Fused Deposition Modeling (FDM). However, the starting material of extrusion-based printing can be both solid and paste (soft) with low viscosity, while the starting material of FDM is wire. In this extrusion-based printing process, material is loaded in extruder (cylinder) before it is extruded through nozzle by ram pressure to create food shape layer – by – layer, as shown in figure 2. The example of food, fabricated via this technique, are dough meat paste and cheese. Lipton, et al. (2010) tested a variety of recipes to print sugar cookies. The result shown that the variation of ingredient concentration effected to fabricate food model especially ratio of butter, yolk and sugar. Therefore, transglutaminase and bacon fat were added to simplify model fabrication. Moreover, Fanli

Yang, et al.[8] applied extrusion at room temperature to print lemon juice gel using the extruder conveying screw, as shown in figure 3. After that, L. Wang, et al. [9] conducted the experiment via the similar system with Fanli Yang, et al. [8] to print fish surimi gel. The results shown that the nozzle diameter, the nozzle movement speed and the extrusion rate affect the quality of 3D food printing, excluding the nozzle height. To print solid stating material, M. Lanaro et al. [10] investigated on melting extrusion for printing complex chocolate model based on machine design, including mechanism design. The results shown that there are two important areas of design in which (1) designing the extruder assembly to be as rigid as possible, thereby reducing flexion and enabling more accurate deposition of chocolate and (2) improving the design of active cooling system to quench the chocolate at lower temperatures, as shown in figure 4.

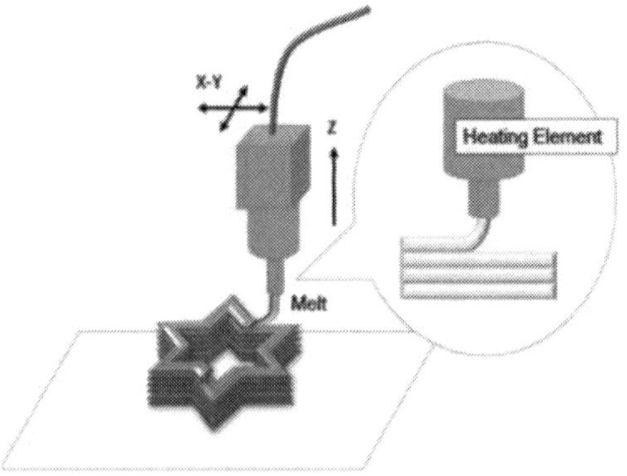

Figure 2. Extrusion-based printing (Peng, Zhou, Jerry, Hong, and Annette, 2015)

Figure 3. The 3D printing technique based on soft-material (A. Anchor, B. Gecko, C. Snowflake, D. Ring, E. Tetrahedron). L. Wang et al. (2018)

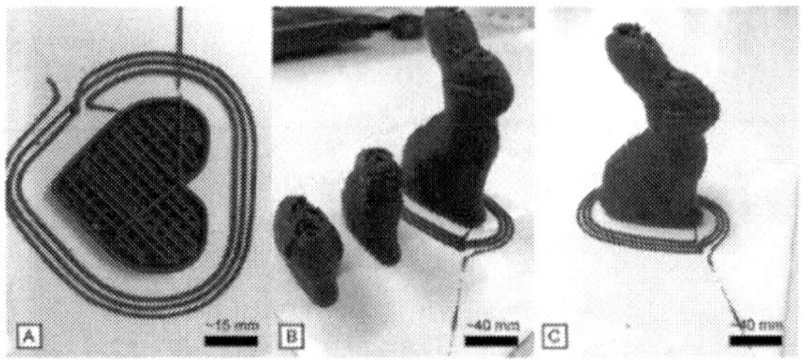

Figure 4. The 3D printing technique based on melting extrusion. (Lanaro, et al. 2017)

2.2 Inkjet Printing (IJP)

The inkjet printing dispenses a material stream of droplets from a thermal head to certain regions for creating the surface filling or decorating on food surfaces, such as cookie, cake, and pizza, as shown in figure 5. This process generally operates by using thermal or piezoelectric heads. In a thermal inkjet printer, the print head is electrically heated to establish pulses of pressure that push droplets from the nozzle [11]. There are two types of inkjet printing methods: a continuous jet printing and a drop-on-demand printing. For the continuous jet printer, an ink is ejected continuously through a piezoelectric crystal by vibrating with a constant frequency. In order to obtain a desired flow ability of the ink, some conductive agents had been added. For a drop-on-demand printer, a valve is a controller ink to eject out from heads under designed pressure. The printing rates of drop-on-demand systems are generally slower than the continuous jet systems, beside the resolution and precision of produced images are higher [12]. The inkjet printer normally handle low viscosity materials; therefore, it does not find application on the construction of complex food structure. Typical deposited materials are chocolate, liquid dough, sugar icing, meat paste, cheese, jams, gels and etc[11].

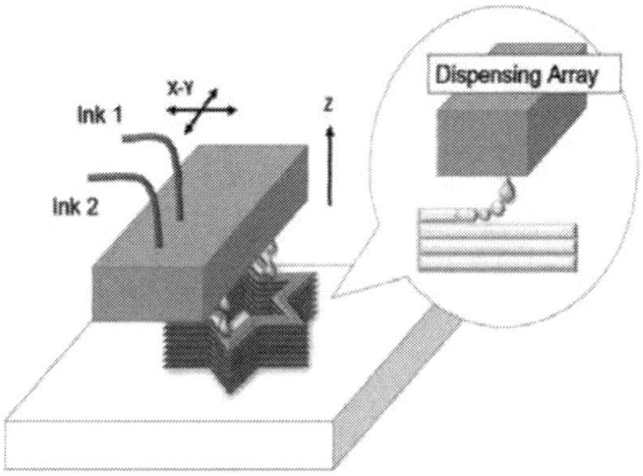

Figure 5. Inkjet printing (IJP) (Peng, Zhou, Jerry, Hong, and Annette, 2015)

2.3 Binder jetting

The binder jetting, which is an additive manufacturing technology, constructs model by using a binder to selectively bond layers of powders. In this process, small droplets of binder with diameters less than 100 μ m are successively deposited on to the powder bed surface, which those are a drop-on-demand print head based on rater scanning pattern. After deposition of the liquid binder, the entire surface of the powder bed is exposed to a fixed amount of heat, which commonly use a heat lamp, for establishing an appropriated mechanical strength via partially cured binder within the generated layer to withstand the shear and gravitational compressive forces involved in the spreading and printing of subsequent layers. These steps are repeated for each layer until the whole feature was completed [13].

For binder jetting process, properties of powdered material and binder are important to the successful fabrication of parts. The binder has to be suitable low viscosity in which surface tension and ink density are suitable properties to prevent spreading from nozzles [14]. S. Holland, et al. were developed food grade inks possessed the necessary properties to be successfully ink jet printed in a Fujifilm Dimatix printer, as shown in figure 6.[15].

Table 1. List of printability 3D printing technology applied for food design.

	Category		
	Extrusion -based printing	**Binder jetting**	**Inkjet printing**
Principle	Extrusion and deposition	Powder binding and binder drop-on demand deposition	Drop-on-demand deposition and Continuous jet printing
Material	Solid-based, Paste material	Liquid-based, Powder-based	Liquid-based, low viscosity material
Processing factor	-Printing head height -Nozzle diameter -Printing rate -Nozzle movement rate	-Printing head types -Printing rate -Nozzle diameter -Layer thickness	-Temperature in print head -Printing rate -Nozzle diameter -Printing head height
Advantage	-Low cost of the entry-level machines -A variety of raw materials are available -Easy to customize	-Large number of potential materials -Very high production speed -Support structures are included automatically in layer fabrication -Low-imaging specific energy -Complex 3D food fabrication	-No waste of model material -High resolution and accuracy -Multiple materials and multiple colors -Fast fabrication
Disadvantage	-Low level of precision and long build time -Unable to build sharp external corners -Anisotropic nature of a printed part -Difficult to hold 3D structures in post-procesing	-Rough or grainy appearance -Post-processing required to remove moisture or improve strength -Limited material -Less nutritious products	-Post-processing may damage thin and small features -Support materials cannot by recycled thus wasted -Simple food design -Only for surface filling or image decoration
Application	Chocolate, Confection, Decorations made of sugar, Candies	Chocolate, Pizza (Powder form), Fake food	Chocolate, Liquid dough, sugar icing, meat paste, cheese, jams, gels
Machine	Choc Creator, AIBOULLY Chocolate, Createbot 3D Food	Chefjet, Fujifilm Dimatix	Foodjet, Filament six-head 3D
Company	Chocedge, AIBOULLY, Createbot	3D systems, Fujifilm Dimatix	De Grood Innovations, TNO

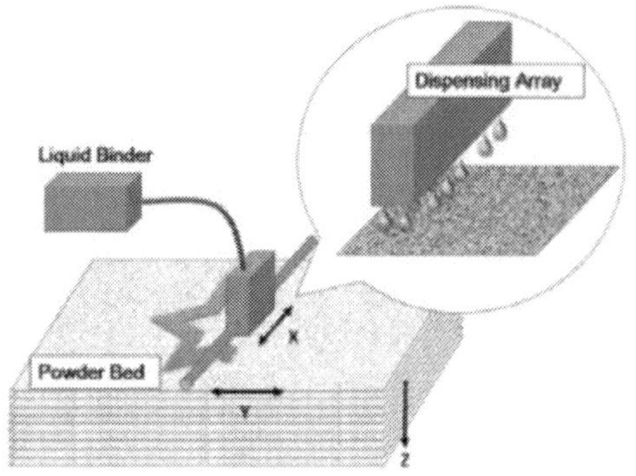

Figure 6. Binder jetting (IJP) (Peng, Zhou, Jerry, Hong, and Annette, 2015)

3 Summary

The Additive manufacturing technologies have been wildly used in several industries including food manufacturing to crate the fantastic food shape. There are variety techniques such as extrusion-based printing, binder jetting and inkjet printing.

In extrusion-based printing, food materials are extruded through nozzle to create food designed shape layer – by – layer. Based on the low viscosity property of material, the suitable materials in this technique are solid-based and paste material. The processing factor of this technique are printing head height, nozzle diameter and nozzle movement rate. The advantage of this technique are the low cost of the entry-level machines, a variety of raw materials are available and easy to customize while the low level of precision and long build time are the disadvantage of extrusion-based printing.

In inkjet printing, food materials are loaded in print head then they is electrically heated to establish pulses of pressure that push droplets from the nozzle. There are two types of printing; a jet printing and a drop-on-demand printing. By using the low viscosity property of material, the suitable materials in this technique is liquid-based. The processing factor of this technique are temperature print head, nozzle diameter and printing rate. The advantages of this technique are high resolution, accuracy and multiple materials while post-processing may damage thin and small features, which is disadvantage of inkjet printing.

In binder jetting, food materials are successively deposited on to the powder bed surface through nozzle. Based on the low viscosity, surface tension and ink density, the suitable material in this technique are liquid-based, powder-based. The processing factor of this technique are printing head type, nozzle diameter and layer thickness. The advantage of this technique are the very high production speed, support structures are included automatically in layer fabrication while the disadvantage are rough or grainy appearance, post-processing required to remove moisture or improve strength of binder jetting.

References

1. M. Annoni , H. Gilberti , M. Strano. (2016). Feasibility Study of an Extrusion-based Direct Metal Additive Manufacturing Technique. 44th Proceedings of the North American Manufacturing Research Institution of SME http:// www.sme.org/namrc. Volume 5, 2016, Pages 916-927.
2. Shannon E. Bakarich , R. Gorkin III , R. Gately , S. Naficy , Marc in het Panhuis , Geoffrey M. Spinks. (2017). 3D printing of tough hydrogel composites with spatially varying materials properties. Additive Manufacturing 14 (2017) 24-30.
3. Teresa F. Wegrzyn , Matt Golding , Richard H. Archer. (2012). Food Layered Manufacture: A new process for constructing solid foods. Trends in Food Science & Technology 27 (2012) 66-72.
4. J.Y. Lee , Jia An , Chee Kai Chua. (2017). Fundamentals and applications of 3D printing for novel materials. Applied Materials Today 7 (2017) 120-133.
5. Nguyen, Tuan C. "Hungry? A Startup Wants You to 3D Print Your Next Meal." http://www.smithsonianmag.com/innovation/hungry-a-startup-wants-you-to-3d-print-your-next-meal-180947702/ (accessed September 18, 2017).
6. Jayakuma, Amrita. "Home-Baked Idea? Nasa mulls 3D printers for food replication." The Guardian.https://www.theguardian.com/technology/2013/jun/04/nasa-3d-printer-space-food (accessed September 18, 2017).
7. New technology often builds on old technology. Understanding how the present printers work and discovering their advantages and disadvantages may help people create more useful food printers in the future.
8. F. Yang, M. Zhang, B. Bhandari, Y. Liud. (2018). Investigation on lemon juice gel as food material for 3D printing and optimization of printing parameters. LWT - Food Science and Technology Volume 87, January 2018, Pages 67-76
9. L. Wang, M. Zhanga, B. Bhandari, C. Yangd. (2018). Investigation on fish surimi gel as promising food material for 3D printing. Journal of Food Engineering Volume 220, March 2018, Pages 101-108
10. M. Lanar, David P.Forrestal, S Scheurer, Damien J.Slinger, S. Liao, S. K.Powell, Maria A.Woodruff. (2017). 3D printing complex chocolate objects: Platform design, optimization and evaluation. Journal of Food Engineering Volume 215, December 2017, Pages 13-22
11. Fernanda C.Godoi, Sangeeta Prakash, Bhesh R.Bhandari. (2016). 3D printing technologies applied for food design: Status and prospects. Journal of Food Engineering Volume 179, June 2016, Pages 44-54

12. Z. Liu, M. Zhang, B. Bhandari, Y. Wang. (2017). 3D printing: Printing precision and application in food sector. Trends in Food Science & Technology Volume 69, Part A, November 2017, Pages 83-94
13. H. Miyanaji, N. Momenzadeh, L. Yanga. (2018). Effect of printing speed on quality of printed parts in Binder Jetting Process. Additive Manufacturing Volume 20, March 2018, Pages 1-10
14. J. Sun, Z. Peng, W. Zhou, Jerry Y.H.Fuh, G. Soon Hong, A. Chiu. (2015). A Review on 3D Printing for Customized Food Fabrication. Procedia Manufacturing Volume 1, 2015, Pages 308-319
15. S. Holland, T. Foster, W. MaNaughtan, C. Tuck. (2018). Design and characterisation of food grade powders and inks for microstructure control using 3D printing. Journal of Food Engineering Volume 220, March 2018, Pages 12-19

A Review of Process Parameters That Effect to Extrusion on Demand

Paphakorn Pitayachaval[1], Thanakhan Baothong[1]

[1]*School of Industrial Engineering, Institute of Engineering, Suranaree University of Technology, Thailand*

Abstract. An additive Manufacturing (AM) process has been introduced to fabricate three-dimension objects from liquid polymers and powder particles e.g. polymers and metals, this process has been also applied to fabricate three-dimension ceramic objects called paste materials which compose of liquid and solid components. To build additive model, the paste material is extruded via pressure force through the nozzle by using ram extrusion, then the model is constructed layer-by-layer on the table to form designed shape. To improve a capability of extrusion process, the Extrusion On Demand (EOD), that refers to the ability of paste extrusion to regulate start and stop, has been developed to improve an accuracy of material deposition. This paper presents a review of process parameters that effect to extrusion on demand such as ram velocity, Dwell time, paste property and extrusion mechanisms.

1 Introduction

Additive manufacturing (AM) is a technology that fabricate three-dimensional (3D) objects by adding material layer-by-layer directly from CAD (Computer Aided Design) model [1]. Traditional AM technologies have been used to fabricate 3D objects based on a raw material states that are liquid-based, solid-based, and powder-based, for example liquid polymers and powder particles including polymers and metals. These materials contains some disadvantages such as high melting point and high cost of raw materials [2]. To solve those disadvantages, paste materials have been applied to fabricate three-dimension parts in additive manufacturing. Pastes material is a kind of material composed of solid and liquid state [3]. Pastes are states that are made up from many different substances and formed by using extrusion method. The paste extrusions have been applied for various industries such as ceramic parts, cosmetic pencils, tiles, food manufacturing, animal feeds, and PTFE wires [4]. Extrusion is carried out by pressing pressure to ram extruder, as illustrated in Figure 1 (a). Normally, the ram extrusion consists of control system and extrusion device. In paste extrusion process, the pressure is generated to force the paste through a nozzle. The paste flow is regulated by controlling the ram movement, which moves down to extrude paste material. The extrudate is deposited layer-by-layer on the table to form designed shape. To improve a capability of extrusion process, Extrusion On Demand (EOD), that refer to the ability to regulate the start and stop of paste extrusion, has be developed to improve material deposition. This paper presents a review of process parameters that effect to extrusion on demand such as ram velocity, Dwell time, paste property and extrusion mechanisms.

2 The paste extrusion process

The Process of paste extrusion is generally carried out in three steps: paste preparation, Forming, and Finishing [3]. In paste preparation, powder components and liquid are mixed in suitable mass proportion to carry out a paste. These components are contained in the barrel. In forming, Pressure is generated to force the paste through a nozzle. Then, the paste is extruded from an extruder by using ram extrusion. The extrudate is deposited layer-by-layer to form a model. During finishing stage, the extrudate is solidified by thermal processing during this stage. Since a conventional extrusion method extrudes the paste by using ram extruder, the extrudate contained head and tail effect as shown in Figure 1 (b). The quality of designed shape depends on these effects. To achieve the accuracy and quality of designed shape, the conventional extrusion method has been improved to control the start and stop extrusion on demand in order to enhance the performance of the extrusion process. In additional, the ram velocity, Dwell time, paste property and extrusion mechanisms are main parameters that influence on the paste extrusion process.

https://doi.org/10.1051/matecconf/201821301013

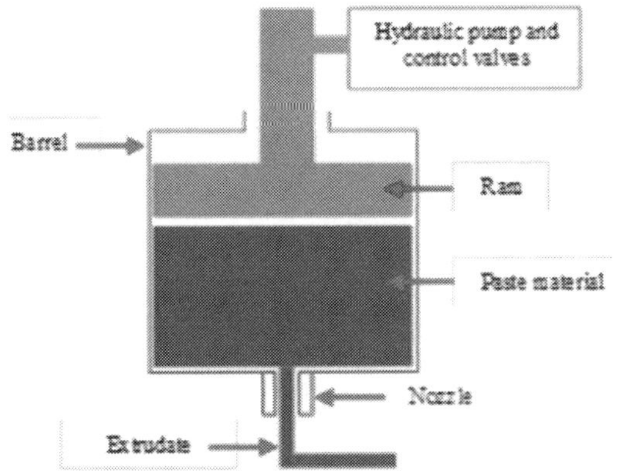

(a) Ram extruder

(b) Head and Tail effect [5]

Figure1. Schematic diagram of paste extrusion for Ram extrusion

2.1 Ram velocity

The paste can flow when the pressure is high enough to force the paste through the nozzle. The pressure force is generated by the ram extruder. The ram is driven by a hydraulic or mechanical system. Because of the relation between ram velocity and pressure force, it is important that the pressure force is generated sufficiently.

2.2 Dwell time

Since there is a delay at start and stop of extrusion, Dwell time is amount of time for start and stop extrusion that process need to compensate the gantry that remains stationary after the force is activated. Determining The Dwell time is important because the excessive Dwell time lead to accumulation of paste material at the start point of printing line. While a few Dwell time cause the nozzle to move forward before the extrusion of paste.

2.3 Paste properties

The paste materials were made up from both liquid and solid compónents, the paste is materials with both a liquid and a solid phase. Paste properties, especially the liquid content and paste viscosity are significant parameter that influences on extrusion process. Variation of liquid contained in paste material cause liquid phase migration that the liquid phase moves faster than the solid particle during extrusion process. So the paste regions of low liquid become drier [6].

2.4 Extrusion mechanism

The ram extruder is widely used for conventional paste extrusion. According to a schematic diagram of ram extrusion as shown in Figure 2 (a), which consists of a ram extruder, barrel and nozzle. The paste material is pressed through a nozzle to generate the deposition process. The start and stop of extrusion is regulated by the ram movement. This conventional extrusion is difficult to fabricate complex shape because of the difficultly of determining time delay for the start and stop of extrusion. So, other mechanisms, used for dispensing technology, has been introduced to improve the extrusion on demand. Needle valve extrusion mechanism consists of a plunger, barrel, a shutter needle and nozzle as shown in Figure 2 (b). The shutter needle was set as a function of the opening and closing valve. Paste material was pressed by the force applied to plunger. Then, the shutter needle was moved up or down by pneumatic force to start and stop of extrusion process. Other, auger valve extrusion mechanism consists of servo motor, barrel and auger valve as shown in Figure 2 (c). This valve was used to extrude paste material by the rotating auger valve. The needle and auger valve have been widely used in dispensing of paste materials [7].

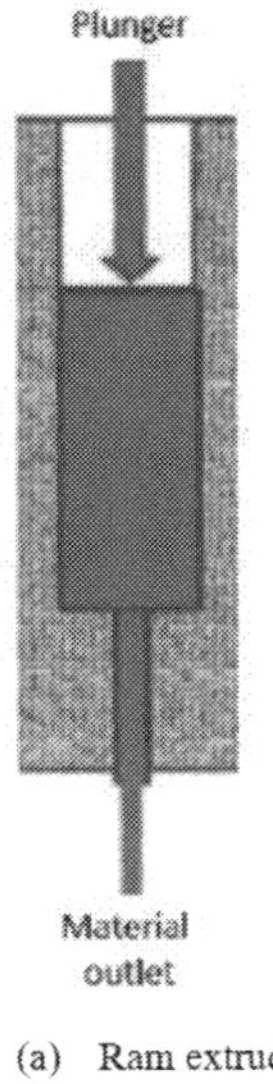

(a) Ram extruder

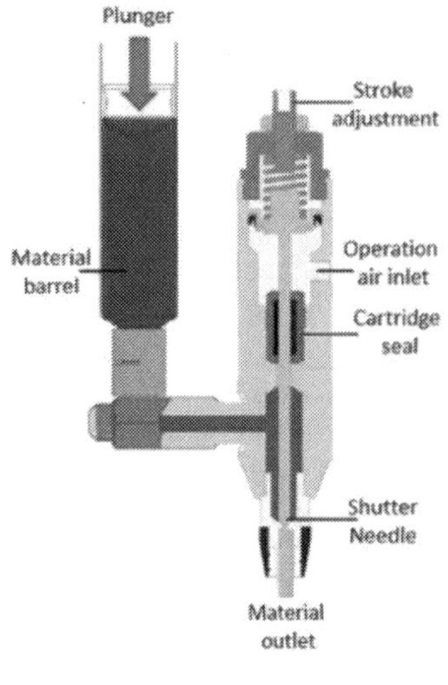

(b) Needle valve

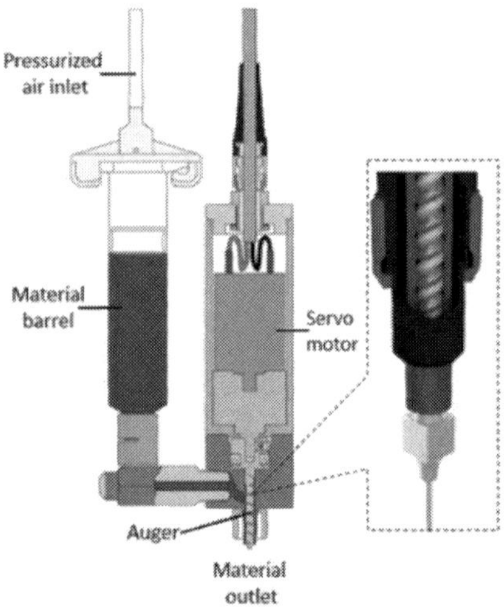

(c) Auger valve

Figure 2. Schematic diagram of ram extruder, needle valve and auger valve extrusion [5]

To improve a capability of extrusion performance, the ram velocity, Dwell time, paste property and extrusion mechanisms are main parameters that have been studies to investigate the effect on the ability extrusion on demand for paste extrusion processes as shown in table 1. Mason et al. (2007) modified the traditional ram extrusion mechanism in order to improve the extrusion on demand. Load cell housing was added to connect directly with the plunger, while plastic syringe was replaced by metal barrel. A new ram extrusion mechanism was implemented to print the line tests with a set of ram retreating velocities (0, 0.25, 0.5, 1, 1.5, 2, 2.5, and 3 mm/s) and extrusion forces of 308N. The results were found that the ram retreating velocities that are higher 2 mm/s able to stop on the extrusion on demand. The influence of the ram velocity on the extrusion force and liquid phase migration in Freeze from Extrusion Fabrication process was studied. Liu and Leu (2009) experimented to investigate liquid phase migration for aqueous Al_2O_3 paste. The aqueous paste and five ram velocities (10, 5, 2, 1.5, 1 μm/s) were used for investigation in this study. The experimental results were shown that highly liquid phase migration was found when ram velocity was lower than 5 μm/s. Oake et al. (2009) present an experiment to investigate the Dwell times that effect to extrusion filament. In this experiment, the Dwell time is set as a function of the reference ram force. This force is applied to activate for paste extrusion. Values experiment are 50%, 55%, 60%, 65% and 70% of the reference ram force 450 N and 475 N. The results were shown that the discontinued filaments were presented at Dwell time less than 65%. In addition, the increasing of the reference ram force from 450 to 475 N was found the accumulation material at the start of filaments. This indicates that a row ram force should be used for extrusion process. Liu et al. (2013) conducted an experiment to study the influence of ram velocity and paste properties on extrusion process by applying series of ram velocities (2,5,10 and 15 μm/s) and three pastes with difference viscosity. The results were found that the extrusion pressure shown an increasing slowly with high velocities (10 and 15 μm/s) whereas the extrusion pressure shown an increasing rapidly at low velocity (2 and 5 μm/s). Therefore, at low velocity, liquid phase was moved into an extrudate. So the remaining paste in the barrel become drier. Drier paste in the barrel need high pressure force to press the paste flow. In addition, the ram velocity of 2 μm/s was experimented to study effect of paste viscosity on liquid phase migration. The results were found that liquid phase migration was found when a higher viscosity paste is used with the lower ram velocity. The paste viscosity had significance effect on extrusion process [12]. The high viscosity of paste materials means a high amount of solid particles. Therefore, high extrusion force had be needed to extrude materials whereas low viscosity paste lead to difficult to form the designed shape. Li et al. (2017) have conducted an experiment to study the influence of extrusion mechanisms, Dwell time and paste property on the start and stop of extrusion. The start Dwell time for ram extrusion, needle valve and auger valve mechanism was 450, 70 and 0 ms respectively. Those mechanisms were used to experiment the ability of extrusion on demand. Dash line printing tests were conducted for all three methods. The line printing tests were printed from right to left and compared the tail and head effect of the printed lines to indicate the capable extrusion start and stop by using the image dash line segments printed. The experimental results were shown that the dash line segments printed by the needle valve and auger valve methods have shorter tails than the ram extrusion method because the start and stop dwell time of the needle valve and auger valve methods are shorter than the ram extrusion method. Then, the filaments with short start Dwell time shown the accuracy of the start and stop of extrusion.

Table 1. Overview of the main extrusion process parameters that effect to extrusion on demand

References	Ram velocity	Dwell time	Past property	Extrusion mechanism
[8]	✓	-	-	-
[9]	✓	-	✓	-
[10]	✓	✓	-	-
[11]	✓	-	✓	-
[5]	-	✓	✓	✓

3 Conclusion

The extrusion methods have been widely used to fabricate 3D objects from paste materials. Those materials are made up from liquid and solid particle. The ram extrusion method is widely used to form paste materials. In this method, pressure force was generated to press paste material through a nozzle by using ram extruder. To improve a capability of extrusion process. The four main parameters were reviewed. The ram velocity, Dwell time, paste property and extrusion mechanisms are those parameters that influence on the paste extrusion process. The ram extruder, needle valve and auger valve based extrusion method were experimented to print the line tests. Because of the shorter Dwell time, needle valve and auger valve shown the ability start and stop of extrusion on demand better than the ram extrusion.

References

1. ASTM Standard F2792-12a. (2012). Standard Terminology for Additive Manufacturing Technologies. ASTM International, West Conshohocken, PA.
2. Lee, J.-Y., J. An and C. K. Chua. (2017). Fundamentals and applications of 3D printing for novel materials. Applied Materials Today 7(Supplement C): 120-133.
3. J. Benbow and J. Bridgwater. (1993). Paste flow and extrusion. Oxford series on advanced manufacturing, Oxford.
4. Amza, C., A. Zapciu and D. Popescu (2017). Paste Extruder-Hardware Add-On for Desktop 3D Printers. Technologies 5(3): 50.
5. Li, W., A. Ghazanfari, M. C. Leu and R. G. Landers., (2017). Extrusion-on-demand methods for high solids loading ceramic paste in freeform extrusion fabrication.Virtual and Physical Prototyping 12(3): 193-205
6. Rough, S. L., J. Bridgwater and D. I. Wilson., (2000). Effects of liquid phase migration on extrusion of microcrystalline cellulose pastes. International Journal of Pharmaceutics 204(1): 117-126.
7. Jianping, L. and G. Deng., (2004). Technology development and basic theory study of fluid dispensing - a review.
8. S. Mason, M., T. Huang, R. G. Landers, M. Leu, G. Hilmas and M. W. Hayes., (2007). Aqueous-Based Extrusion Fabrication of Ceramics on Demand.
9. Liu, H. and M. C. LEU., (2009). Liquid Phase Migration In Extrusion Of Aqueous Alumina Paste For Freeze-Form Extrusion Fabrication. International Journal of Modern Physics B 23(06n07): 1861-1866.
10. Oakes, T., P. Kulkarni, R. G Landers and M. Leu., (2009). Development of extrusion-on-demand for ceramic Freeze-form Extrusion Fabrication.
11. Liu, H., J. Liu, M. C. Leu, R. Landers and T. Huang (2013). Factors influencing paste extrusion pressure and liquid content of extrudate in freeze-form extrusion fabrication. The International Journal of Advanced Manufacturing Technology 67(1): 899-906.
12. Hong, S., C. Sanchez, H. Du and N. Kim., (2015). Fabrication of 3D Printed Metal Structures by Use of High-Viscosity Cu Paste and a Screw Extruder. Journal of Electronic Materials 44(3): 836-841.

A High Degree of Cleanliness of the Electric Air Supply Valve Design

Luo Hu

Mechanical and Information Engineering Department, China University of Mining and Technology, 100083, Beijing, China

Abstract. This paper discuss an new type servo valve, aiming at two main defects of performance of globe valve which are commonly used in mine safety cabins in China, in order to bring up a better solution to replace old valves. On one hand, most of globe valves are manual ball valves whose oxygen content cannot be accurately controlled and can only be estimated by human respiratory sense, ended in large errors[1]. On the other hand, although in market there is oxygen-concentration test equipment, the separation of valve components and manual control switch, leads to complicated procedures and management inconvenience. For solving these problems, an innovative valve mechanical structure is designed, with its servo system bringing up. The proportional valve measures the room oxygen concentration in real time, closed-loop controlling of electric valve-opening, so that the concentration of oxygen in the ambient air is maintained at a desired level. The principle of this implementation is that an explosion-proof motor drives the roller cam to convert valve core into reciprocating linear motion, separating O-ring and valve body end-face to control valve core achieving the opening and closing functions. Since there is no relative motion between O-ring and valve body, it definitely avoids problems of uncontrollable particles caused by abrasion and ensures a high cleanliness of flow path. In addition, electric servo valve controls the precise oxygen concentration, automatically. Compared with manual air supply valve, this valve has advantages of high cleanliness, reliable sealing, easy assembly, maintenance, and automation. It provides a new technical approach for design of special industrial and mining value.

1 Introduction

In cases of dangerous situations such as coal mines and other special occasions during operation, to ensure the safety of workers, it needs to set up an escape safety cabin for temporary sheltering and waiting for rescue. In its life support system, the Oxygen and Nitrogen gas supply lines need supply valves to maintain the gas balance in the cabin. Manual valve switches do not have the function of automatic maintaining the gas concentration balance. The oxygen concentration can only be determined by human breath[2], ended in big error of Oxygen concentration. And the oxygen concentration may become too low to breath or too dense to breath. At the same time, when the number of valves gets too large, opening and closing the manual valve one by one is not only time-consuming but also more prone to safety accidents. Therefore, it is necessary to propose an electric servo air supply valve[3].

The structure of the current valves are main two kinds: filling ball valve and plug valve. For filling ball valve, although its seal property is reliable, at the switch seal interface, close contact with each other relative rotational friction, resulting in wear and tear, excessive wear debris, affect gas cleanliness requirements, while human respiratory system gets a detrimental effect. Plug valves have been widely and successfully used in the pneumatic and hydraulic industries owing to its oil's reduced friction. However, due to the absence of a lubricant in a high-purity gas environment, there is a possibility of the O-ring being cut, causing a sudden leak failure, which leads this kind of valve to be used less and less. Diaphragm shut-off valve can control the excess material generated, but it`s not suitable for escape chamber small flow of gas delivery occasions[4].

With the gradual improvement of the cleanliness standard, the original cut-off valve is difficult to meet the high cleanliness and automation requirements. Therefore, it is urgently needed to develop a new type of servo valve to overcome the above insufficiency of the gas supply valve. The proposed new type of electric gas supply servo valve can realize the sealing function by compressing and separating movement between the sealing interfaces to prevent the generation of particles. High cleanliness features, automatic opening and closing and high reliability to maintain the balance of gas supply.

2 New fast opening and closing valve design

2.1 Electric supply valve technical indicators and technical requirements

According to the general requirements of the safety cabin pipeline conditions, the main technical indicators are shown in Table 1.

Technical requirements: This is a low-pressure gas supply oxygen valve, and has a quick opening and closing function, opening and closing the frequency of moderate. As the gas supply to the body to breathe, so the supply of gas is not allowed to produce more than 0.1mm particles to meet the cleanliness standards. Due to the output of pure oxygen gas, there is a mandatory restriction on the gas flow rate for safety reasons. The flow path surface must have inert oxygen[5]. The technical route is shown in Figure 1.

2.2 Electrical oxygen valve technology principles and characteristics

2.1.1 Drive components technical principle

Driven by motor reducer Eccentric cam guide rod and the spool reciprocating linear motion, when the eccentric cam, the valve ring to move the valve ring to separate the valve body to open, as shown in Figure 1.When the reducer shaft is rotated 180 degrees, it is the maximum open position, and the valve is closed when rotated 180 degrees again.

2.1.2 The original motivation features

The prime mover adopts a motor driven by a reducer rather than a direct motor. On the one hand, the reducer increases the operating torque so that a smaller motor can drive it. On the other hand, it is not easy to slip when the output shaft has a large torque when the device is not running, it can effectively prevent it from inverting.

Table 1. Electric oxygen valve technical indicators

Physical quantity	**Symbol**	**Value (unit)**
Work pressure	P	0.1MPa
Max-pressure	P_{max}	0.5MPa
Output flow	Q_1	$100 \sim 400ml/min$
Internal leakage rate	q_1	$\leq 10^{-3}$ Pa·L/s
Leakage rate	q_o	$\leq 10^{-4}$ Pa·L/s
Control method	/	S/W
Rated current	I	$1 \sim 2A$
Rated voltage	V	$12V$
Medium	/	Oxygen, nitrogen
Switching time	t	$< 1s$

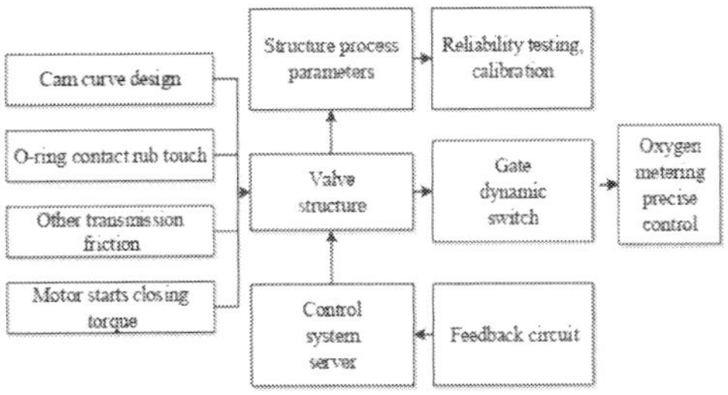

Figure 1. Technical route

2.1.3 Contact seal features

Through the sealed interface in situ movement between the compression and separation to achieve sealing function, no relative rotation between the seal vice or sliding to prevent the generation of particles to meet the high cleanliness characteristics.

2.1.4 The reliable sealing pair O-ring features

The O-ring 1 is fitted on the guide rod and the end face of the valve body to form a sealing pair, the sealing pair interface only exists reciprocating movement, the contact process is the O-ring compression deformation process, the state is the same as the static seal on the face. To overcome the plug valve rotation process rotary seal with the filler ball contact rotating friction and wear bring particulate matter, which is shown in Figure 2a and 2b.

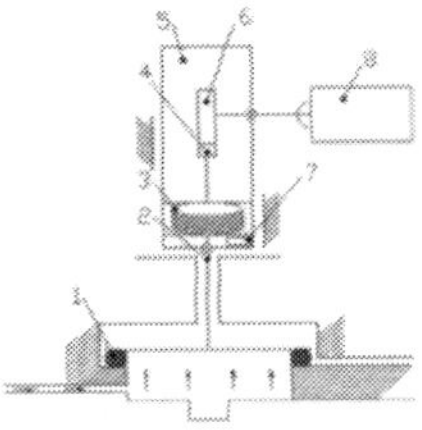

Figure 2a. Electric valve flow path close state

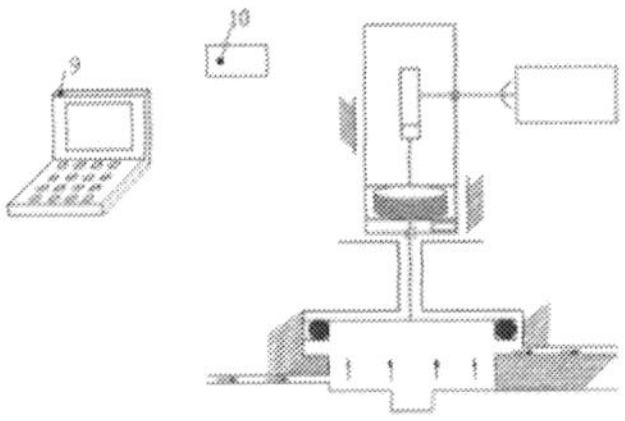

Figure 2b. Electric valve open

1: Sealing ring 2: Guide shaft 3: Disc spring 4: Roller 5: Valve body 6: Groove
7: Dial 8: Motor 9: PC 10: Oxygen detector

2.1.5 Seal compression amount and size chain compensation

The amount of O-ring compression is a key parameter in this design. The end seal compression ratio is generally 22%~26%, which determines the compression range. Using the principle of the shortest dimension chain to control the compression rate can greatly reduce the processing cost. As shown in Figure 3, the distance between the k-plane of the valve core and the m-plane of the valve body is the amount of compression and the increasing ring L3 and the reduction ring L2 constitute the smallest size chain. Only controlled by two size chains easy to guarantee, which is shown in Fig3. To ensure the value of the design choice eccentricity e is greater than the amount of compression; that e =(0.2mm~0.3mm), while avoiding rigid transport interference, the valve core and the guide rod to increase the stiffness of the pressure Spring 2, the stiffness of the valve core seal compression stiffness of 5 to 8 times. This avoids interference with the rigid body and reduces machining accuracy while also ensuring the amount of compression required for the seal[6].

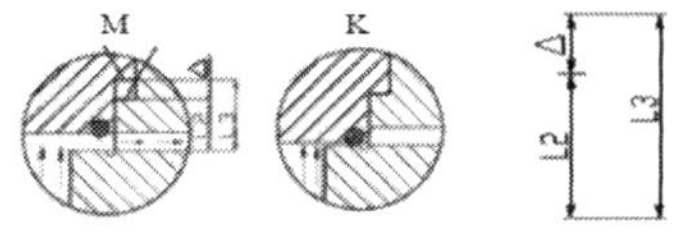

Figure 3. Ring size chain diagram
L1: Seal the amount of compression; L2: body step size; L3: spool step size

2.1.6 System action status

The system operation status is as shown in Figure 3 and the system passes through the oxygen concentration detection sensor in the mine environment, oxygen concentration data monitoring, feedback data, real-time and target oxygen concentration contrast so poor, the deviation of the valve to open and close the size of the dynamic control at the same time, the system designed by the valve high The accuracy of the linear potentiometer to promote the displacement of the last level of the spool to measure the known oxygen content of the spool displacement of the worst comparison, the system of these two deviations after the integration of the valve opening and closing control, In order to achieve a more accurate control of the ambient oxygen content, which is shown in Figure 4.

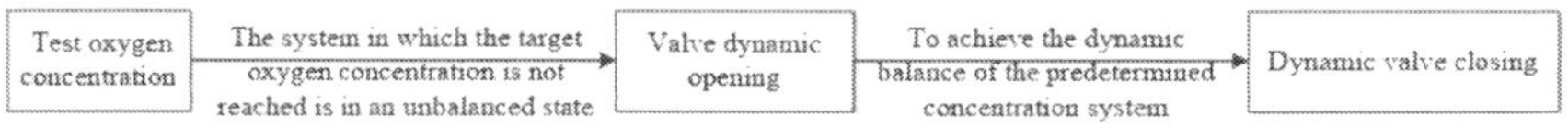

Figure 4. System operation status

3 Calculation and check

3.1 Motor reducer starting torque calculation

This chapter established the oxygen shut-off valve statics model shown in Fig 5. In the establishment of its static model, consider the rolling contact friction cam, guide rod sliding friction, gas force and seal pressing force. The pressing force of the seal ring is the critical boundary condition to be calculated. The relationship between the compression amount and the pressing force of the rubber seal ring is obtained through the finite element method[7]. The relationship between the rotational resistance torque and the working resistance torque is as follows:

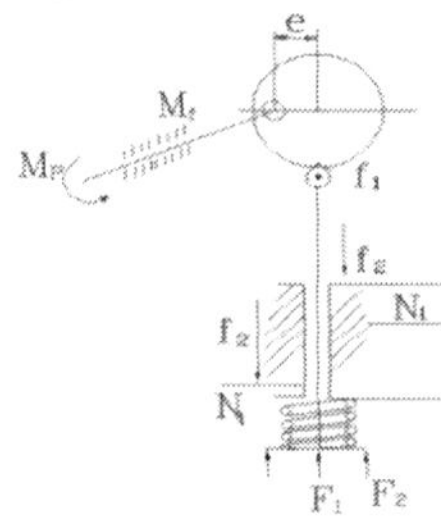

Figure 5. Oxygen servo driving mechanism static model

Table 2. Oxygen servo valve transmission parameter

Symbol	Physical meaning	Value(unit)
e	Cam eccentricity	3mm
r	Cam radius	12mm
P(max)	Cam seals to withstand the maximum pressure	2Mpa
d	Shaft diameter	10mm
L	Guide rod length	60mm
μ_1	Roller cam friction coefficient	0.05
μ_2	Guide rod and sliding friction coefficient	0.3
μ_3	Guide rod and valve body friction coefficient	0.3

3.2 Relationship between O-ring-compression and compression pressing force

The rubber O-ring was selected to have a Shore hardness of 75 degrees and was treated with a large elastic deformation. The Mooney model was selected for the material properties to estimate using the finite element contact analysis[8]. The two sides of the O-ring were rigid geometrical boundaries and 1/4 Meshing as an axisymmetric analysis model, shown in Figure 6a, the millimeter unit system, O-ring cross-section diameter of 2.65mm, an inner diameter of 8mm, the amount of compression 0~1 mm, obtained Cauchy stress during compression, contact compression and compression Relationship between. Fig6b Cauchy equivalent cloud chart, Figure 5c 25% compression rate compression force 70N.

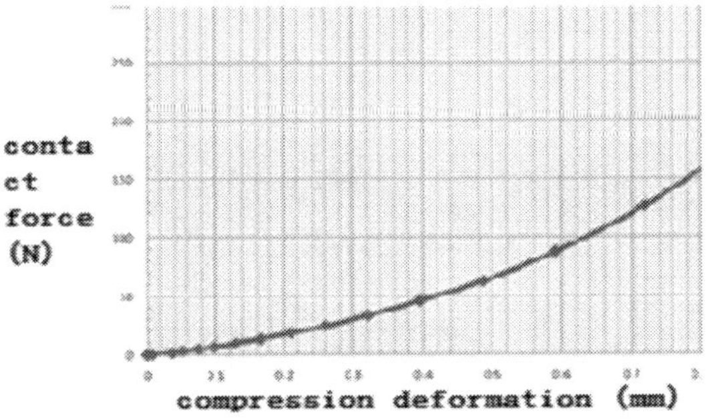

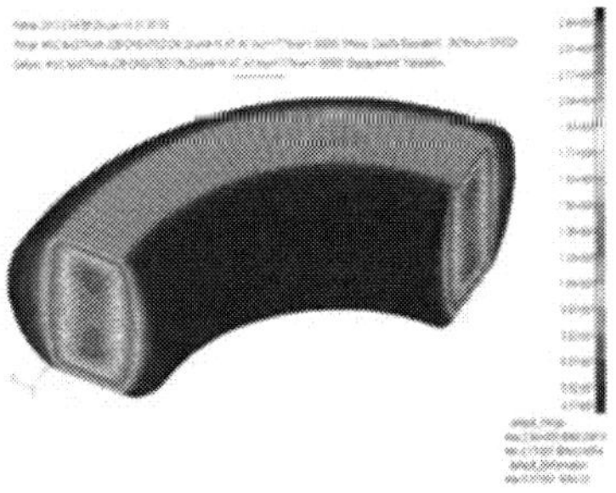

Figure 6a. Cauchy stress ring

Figure 6b. Contact force - compression deformation curve

Calculated by the handle open torque 0.4Nm, here to 0.8Nm, if the reducer gear ratio to take 3.6, the motor output torque 0.22, you can choose 28 stepper motor, if the spool reciprocating frequency of 1, the motor speed 3.6r / s.

Table 3. Torque calculation formula

Symbol	Physical meaning	Expression	Value	Unit
F_1	Gas force	$F_l = \pi d^2 P_{max} / 4$	39	N
F	Disc spring max-pressure	$F_2 = k\Delta$	20	N
/	Sealing force	Finite element analysis results	70	N
f_1	Cam roller friction	$\mu_1(F_1 + F_2 + f_2)$	3	N
F_2	Eccentric wheel above guide rod normal force	$F_1 + F_2 + f_1 + f_2$	119	N
f_2	Guide sliding friction	$f_2 = 2\mu_3 F_2 \cdot e / L$	7	N
M_h	Reducer output torque	$F_2 \cdot e$	0.4	Nm
N	Motor speed	$f \cdot i$	/	R/s

3.3 Cam parameters design and check

Cam and roller contact (cam pair) at the normal load, in units of N. Structure operation diagram is as shown as Figure 7, simplified model of cam transmission.

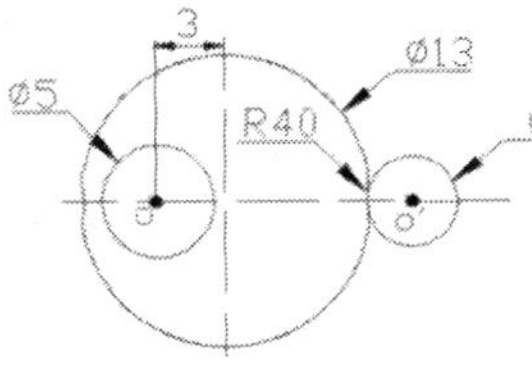

Figure 7. Simplified model of cam transmission

$$\sigma_H = \frac{F}{\pi L} \sqrt{\frac{\frac{1}{\rho_1} + \frac{1}{\rho_2}}{\frac{1-\mu^2_1}{E_1} + \frac{1-\mu^2_2}{E_2}}} \quad (1)$$

ρ_1、ρ_2: The radius of curvature of the actual profile of the cam and the roller radius in mm.

L: Cam and roller contact line length.

μ_1、μ_2: Poisson's ratio of cam and roller material, respectively.

E_1、E_2: Respectively elastic modulus of cam and roller material quantity MPa.

Where, b = 0.005mm, F = 112N, σ_H=150N is calculated 108MPa, less than the allowable stress to meet the design requirements.

Curves of the movement of the law to determine: As the cam lift away from the displacement $x = x_0 + e \times \cos\theta$, the sealing ring diameter of 2.67mm, the particle center x_0 from the valve body end is 1.34mm, but due to the compression rate of 15% The body end surface is 1.12mm, the cam is eccentric from the nearest minute groove with the shortest farthest to prevent the motor from starting is free to reverse[9].

3.4 Servo valve oxygen flow calculation

Assuming that the motor speed is n r/m, the radial R_T mm, displacement of the guide shaft is related to the time:

$$s = R_T + e\sin\varphi t = R_T + e\sin\frac{2\pi n}{60}t \quad (2)$$

https://doi.org/10.1051/matecconf/201821301014

It beginning of the rising phase of the guide shaft, the oxygen outflow area is shown in Figure 8 below: where 1 is the outlet of the valve and 2 is the end surface of the guide shaft.

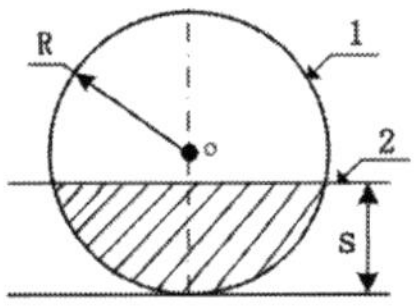

Figure 8. Radial displacement of guide shaft and ventilation cross section

4 Oxygen servo valve structure design

4.1 Structure design

Oxygen supply structure of the valve as shown in Figure 9, the valve body as a whole structure, the bottom of the extension of the installation interface, that is, with a hole for the valve fixed, and the base vertical holes above the pipeline connection with the entrance and exit. Handle and reducer 11 layouts above the valve body, and the valve body perpendicular to the handle and valve body are dust cover. The valve core 5 is designed as a fitting groove structure, and a disc spring 6 is embedded between the valve core 5 and the guide shaft 7 through the fitting of the fitting groove so as to avoid the rigid motion interference. The disc spring 6 is provided with a radial groove on the outer circumference to release the rigidity and increase the deformation of the disc spring. The action of the stop block prevents the disc spring from becoming unstable and can reduce the precision of the sleeve and the guide bar. O-ring 2 will be fixed by spinning the valve dovetail groove to prevent its fall off, and then through the reducer output shaft rotation to achieve the valve switch. Open the valve overall height of 174mm, 114mm center distance shown in Figure 9, to achieve the valve lightweight design.

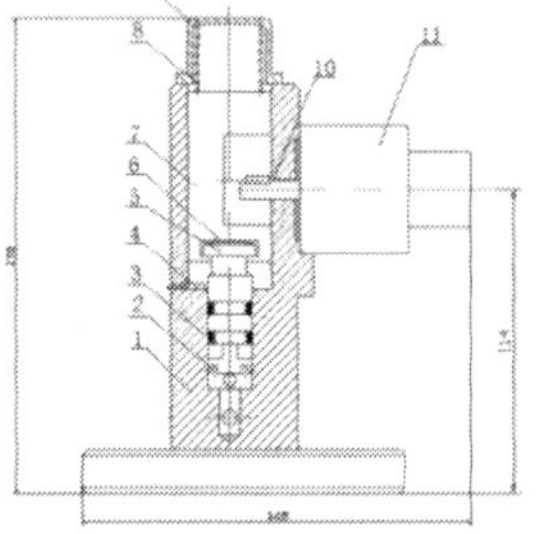

Figure 9. Oxygen servo valve structure

1: Body 2: Seal ring 3: Secondary seal 4: Linear potentiometer 5: Spool.6: Disc spring 7: Guide shaft 8: Spring 9: Bonnet 10: Gear unit output shaft 11: Gear unit

5 Valve servo system

5.1 Servo valve controllable range

To find out if the valve end is in contact with the valve body and achieve a 15% compression ratio to maintain seal reliability, introduce a linear potentiometer to measure spool travel distance as shown in the dimension chain diagram in Figure 10.

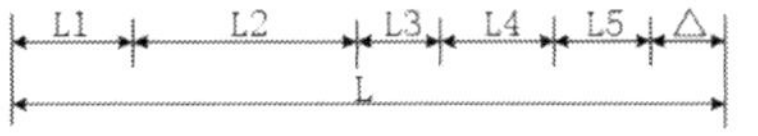

Figure 10. Schematic test system

L1: ring compression, L2: the length of the valve body and the length of the valve body, L3 disc spring length, L4 and the sum of the radius of the roller and the L5 potentiometer distance Δ for machining errors between parts, calculated L5 is within the range of (2 ± 0.2) mm and (8± 0.2) mm, but for practical verification and better control performance, a linear potentiometer is introduced to measure the reciprocating movement of the spool.

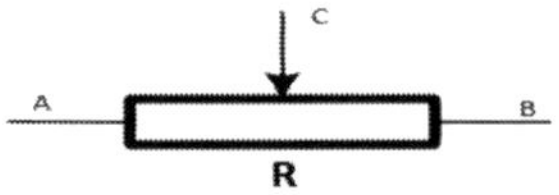

Figure 11. High precision linear potentiometer circuit model

Linear potentiometer installation design shown in Figure 9, at 4th position; Linear potentiometer test principle shown in Figure 11, A and B are respectively connected to the reference voltage positive and negative, C-terminal angle simulation voltage output.

The corresponding relationship between the voltage signal of the linear potentiometer sensor and the linear displacement signal is (3), where the line displacement of the C port to ground represents the output value of the analog voltage signal, indicating the linear displacement of the linear potentiometer, that is, the displacement of the spool, Corresponding to the position of the C sliding terminal on the resistor R. The proportional coefficient of the linear potentiometer is determined by the relation (4), is the ratio of the reference voltage applied to the sensor to the maximum detectable line distance.

$$V_{out} = K_p L \tag{3}$$

$$K_L = V_{ref} / L_{\max} \tag{4}$$

The C-terminal analog voltage directly to the Atmel company Atmega-8-bit microcontroller analog digital AD interface, the AD precision of the microcontroller is 10, read the analog voltage value through the conversion relationship (5) converted to obtain spool displacement, that is realized with a linear potentiometer reciprocating distance measurement.

$$L = AD \cdot L_{\max} / 2^{10} \tag{5}$$

The range of values measured by the analysis of Figure 11 is: $L = AD \cdot L_{max} / 2^{10} = (2\pm0.2, 8\pm0.2)\,\text{mm}$.

Real-time monitoring of the spool through the actual distance from the actual concentration of oxygen content in the room, two quantities at the same time return back to the Atmeg8 microcontroller; Then the controller will make data, which is feedback by the dual potentiometer, integrated to give out command of motor controlling signal to drive the angle rotating system.

5.2 Servo control system

For accurate control of indoor oxygen concentration, the design of double closed-loop control system, the system input target oxygen concentration of $R_{in}(s)$ Drive amplification circuit amplification factor is k_p; Permanent magnet brush motor transfer function is $G_0(s)$; Reducer reduction ratio k_i; Groove roller transfer function is $G_1(s)$, the output displacement of x_n; Displacement x_p and oxygen concentration $Q_{out}(s)$ between the transfer function $G_2(s)$; System output oxygen flow $Q_{out}(s)$; And indoor oxygen concentration of $R_{out}(s)$ between the relationship of $G_3(s)$; The groove wheel output displacement x_p and indoor oxygen concentration $R_{out}(s)$ feedback to the controller at the same time, the controller algorithm data fusion system output control, dynamic adjustment valve opening and closing state, in order to achieve better results. That is, the transfer function and block diagram of the control system step by step are shown in Figure 12.

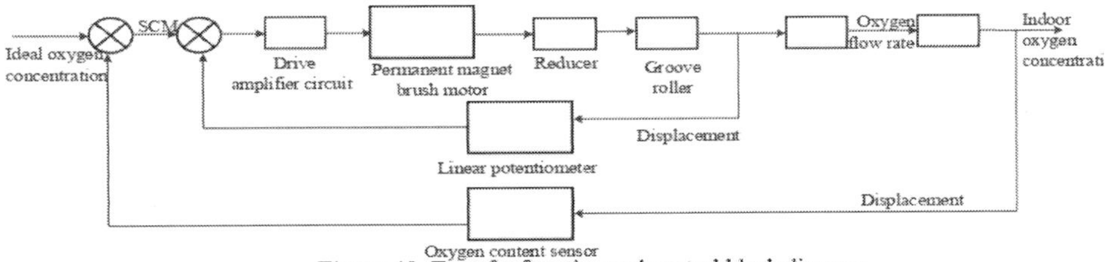

Figure 12. Transfer function and control block diagram

Wherein, x_p represents the transfer function between the output displacement x_p of the grooved roller and the oxygen concentration $Q_{out}(s)$, which is also the expression of the core of this design. Fitting through multiple simulation experiments, the optimal fitting expression (6) as follows:

$$G_2(s) = \frac{Q_{out}(s)}{S_{ot}(s)} = \frac{0.232s + 5}{0.8s^2 + 2.765s + 7.762} \qquad (6)$$

5.3 Simulation & test

Through the system control block diagram of the above design, the simulation experiment is carried out by MATLAB, and the displacement curve of the output shaft rotation angle and seal ring axial calculation and test is shown in Figure 13. The relationship between the output oxygen flow and the spool displacement is shown in Figure 14.

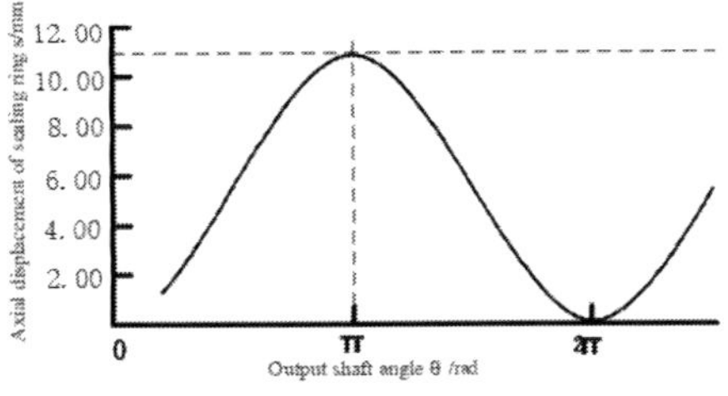

Figure 13. Output angle with seal displacement relationship

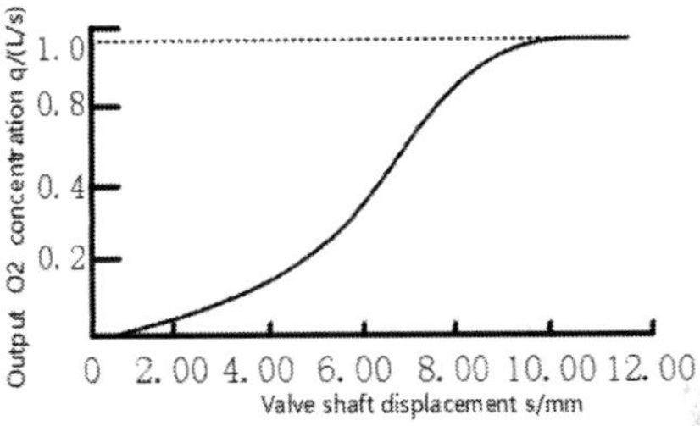

Figure 14. Output oxygen flow vs. valve core displacement

Analysis of the relationship between the spool displacement and the relative drive shaft rotation angle and the output oxygen flow in Figure 13 and Figure 14, shows that there is a nonlinear relationship between the system valve shaft displacement and the output O_2 concentration; To test this system under the action of the step target input signal, with 2 cubic meters of limited indoor space to simulate the test of oxygen concentration in the air; As shown in Figure 15, the flow of oxygen output to the indoor environment is a point-source space diffusion of the $u(\vec{r}, t)$ function of time t and space $\vec{r} = (x_1, x_2, x_3)$, where analytical solution satisfies the relation (7), where M_0 is the initial conditions and M is concentration coefficient.

$$Q = -\frac{1}{r} e^{-\frac{r}{t}t} M + M_0 \qquad (7)$$

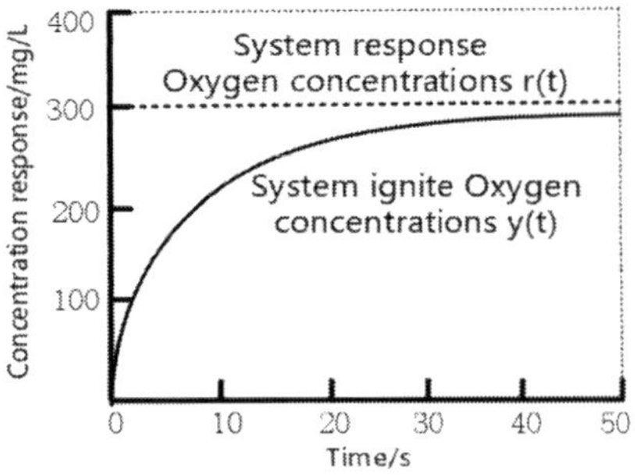

Figure 15. Step target oxygen concentration response curve

According to the normal oxygen content in the air, that is, the oxygen content under the standard condition is 310mg / L, in the limited space of $2m^3$, the system set the target oxygen concentration to 300mg / L, after t = 260s, the system space basically reached the goal Oxygen concentration. Simulation results show that, in line with the actual situation, to meet the design requirements.

6 Conclusions

In this dissertation, we improve the design and feasibility of the existing mainstream oxygen supply shut-off valve in China and carry out the simulation of the control system to conclude that the oxygen concentration of the valve design of the valve is in line with the reality, which can provide powerful force for the mine and safety valve system Of the equipment reference; I would like to emphasize the valve guide shaft reciprocating seal and end double seal, "two protective measures" to ensure that its oxygen does not leak, high reliability features. In the innovative structural design of the valve system by adding double closed-loop feedback control, making the valve opening and oxygen flow output is absolutely controllable so that the concentration of oxygen in the final confined space to be high-precision control, the control process is fully automated man-made Operational inconvenience and difficult to control and other issues are well resolved, that the valve to facilitate management, eliminating the need to regulate the manual valve operation, precisely control the room oxygen content, better performance, has good prospects.

References

1. Xiang gui-sheng, Labor Protection [J]. Emergency Refuge Chamber In The Application of The Mine Disaster Rescue 2006, (9); 92-93.
2. Xue-Guan SONG, Lin WANG, Analysis and Optimization of Rubber Sealing Mechanism Of Ball Valve [J]. Transactions Of Nonferrous Metals Society Of China 2009, (19):220-224.
3. Wang sheng, Mine Emergency Rescue Capsule Technology Abroad [J]. Journal Of Safety Science and Technology **8**(2010), 42-46.
4. Zhang Bao-sheng, ChenJia-qing, Sheet Metal Floating Seal Finite Element In Model Study [J] China Petroleum Machinery **6**(2011).
5. Cai Yi-gang, Theory Analysis Of Fluid Transmission Line Dynamics. Fluid Transmission Line Dynamics [M] Zhejiang University.
6. Chen Cheng-jun, Combining Face Seal Performance Oriented Assembly Connection Process Design[J] Journal of Xi'an Jiao Tong University J Xi'an Jiao Tong University **46**(2012).
7. Qi Lin, Test Study on the preparation of a Variety of Process Modification Layers On The Surface Of Valve Materials [D]. Master Thesis of Shenyang University of Technology (2012).
8. Lv Pei-wen, Parameters of Practical Valve Design. Practical Valve Design Manual[M]. China Machine PRESS.2012
9. Partial differential equation with boundary with Fourier series and boundary value problems[M] China Machine PRESS.2007

Characterization of Mortar from Church Ruins in Barangay Budiao, Daraga, Albay

John Carlo A. Mangay[1], Eric T. Miranda[1], John Mark S. Anicas[1], Angel S. Recto[2], Jan-Michael C. Cayme[1]

[1]*School of Chemical, Biological, Materials Engineering and Sciences, Mapúa University, Muralla St., Manila, Philippines, 1002* [2]*College of Social Sciences and Philosophy, Bulacan State University, Malolos, Bulacan, Philippines, 3000*

Abstract. The lack of information about the correct ways of restoring historical masonry materials from cultural heritage structures tends to create more damage to the said structures. According to Cesare Brandi's theory of restoration, existing historical materials must be replaced with their equivalent. This paper presents the chemical composition of historical mortars acquired from church ruins of barangay Budiao, Daraga, Albay, Philippines. The historical mortar sample was characterized using petrographic analysis, sieve analysis, atomic absorption spectroscopy (AAS), thermogravimetric analysis (TGA), and infrared spectroscopy (IR). Petrographic analysis shows that the mortar sample is bounded by 90% organic material particularly coral and is held together by microcrystalline calcite (micrite) matrix that comprises of 10% of the total sample mortar composition. Sieve analysis shows a well-graded particle distribution. Atomic absorption spectroscopy (AAS) was also utilized to determine the %Ca for each sieve fraction using HCl for acid digestion. TGA thermograph shows the non-linear drop in weight between 740°C and 850 °C, which corresponds to the degradation temperature of calcium carbonate or limestone. IR analysis shows essential characteristic peaks of $CaCO_3$ at 712 cm^{-1}, 874 cm^{-1}, and 1437 cm^{-1}.

1 Introduction

Just about two kilometers away from the bell tower of Cagsawa Church are the ruins of the town of Budiao, another place buried by the lava flow that came from Mayon Volcano, that fateful day of February 1, 1814. Only a very small section of the church protrudes above ground and on the edge of a village road. The town of Budiao gained popularity during the Spanish colonization period because it was there that the foreign colonizers set up an arena for bullfights [1]. On the other hand, historical buildings and National Cultural treasures have an important role of telling the story of what happened in the past in a certain place or society. History taken from these old buildings can be utilized as a vital source of information in numerous disciplines of arts and sciences, also historical buildings are considered to be an identity of a place or society (National Cultural Landmarks). For renovators and conservators, it is important for them to understand the characteristics and properties of the materials used in constructing these pieces of historical buildings and through understanding it, producing a compatible substitute material with the same composition as the original mortar is essential in restoration and conservation works. Furthermore, this study can be applied to other fields of disciplines like architecture, history, archeology, anthropology, etc.

The usual components of a historical mortar are composed of the following: the matrix or binder which is usually calcium carbonate from limestones, and the dispersed medium or the aggregates which is usually sand, the third component of a historical mortar comprises of the least fraction among the components is the additives, which are usually made up of clays. Lime based mortar technology were used in the Roman times until late XVIII century, it is when they replaced hydraulic binders to Portland cement as substitute binders and render, by which completely utilized in civil and military construction. By that time, the insufficient involvement in historic structures with systematic modern solutions lead to the disappearance of original renderings and finishing, which created new problems regarding compatibility issues due to the differences in mechanical and physical properties, which includes strength, porosity, permeability, color, and of course chemical composition [2].

Conservation and renovation of historical sites and buildings is a big factor in preserving a nation's history. According to Cesare Brandi's theory of compatibility, existing historical materials must be replaced with their equivalent. Hence, proper methods and materials must be used in order to preserve/restore/conserve a cultural heritage structure. Poor restoration techniques and lack of information about these construction materials are being used in the Philippines, like using Portland cement as reconstruction material to replace the original mortar and also as renderings in cultural heritage structures.

The main focus of this study is to characterize a sample mortar acquired from the church ruins of Budiao, Daraga, Albay, which can be used as a baseline data for manufacturing a compatible substitute material for the renovation and preservation of the church ruins. Furthermore, this research provides useful analytical techniques in identifying the components of a historical mortar.

2 Experimental Methods

2.1 Sampling

Sampling is an essential process that will ensure the success of the study. The historical mortar sample was acquired in November 2016. A loose brick with its mortar still attached is where the historical mortar sample originated which will be used for later physical and chemical testing. The historical mortar samples were carefully detached from the loose brick using a stainless-steel chisel and hammer. The detached mortar samples are then placed in zip lock containers and then labelled accordingly.

2.2 Analysis

2.2.1 Petrographic Analysis

Optical microscopy of thin sections of mortar samples will be used for the petrographical-mineralogical characterization of the sample mortar's components [3]. This technique will allow to have information on the mineral constituents of the mortar sample, as well as to determine the binder matrix to the aggregates.

2.2.2 Sieve Analysis

Around 5.76 g of historical mortar sample was deliberately disaggregated and gently crushed using a ceramic mortar and pestle.To acquire information about the grain size distribution of the resulting material, sieve analysis was utilized [3], using USA Standard Testing Sieve ASTME-11 Specification, with sieve sizes of 4.75, 2.36, 1.18, 0.600, 0.425, 0.250, 0.150, and 0.075 mm. The lowest sieve size which is on the "pan" (<0.07 mm) will also be considered, and each sieve fraction will be named according to the size of the sieve by which it was retained.

2.2.3 Atomic Absorption Spectroscopy

Atomic absorption spectroscopy allows the quantification of elemental composition of the sample mortar specifically its calcium content. Each retained sample from each sieve size were analyzed via AAS digested in hydrochloric acid (HCl).

2.2.4 Thermal analysis by thermogravimetric analyzer (TGA)

A 6.50 mg of dry, unsieved historical mortar sample was heated within the temperature range of 25 °Cto 900 °C at 10 °C/minute heating rate in dynamic air. The thermograph is acquired using a Mettler Toledo thermogravimetric analyzer. The weight loss of the historical mortar is plotted as the temperature increases.

2.2.5 Infrared (IR) spectroscopy characterization

A Thermo Scientific Nicolet 6700 FT-IR Spectrophotometer in the mid frequency region was utilized to record the infrared spectrum. A mixture of historical mortar sample and KBr powder were grounded and compacted to produce a transparent, thin pellet. Initially, the sample mortar was heated at 100 °C in an oven for 3 hours to guaranty that the sample is completely dry.

3 Results and Discussion

3.1 Petrographic analysis

Petrographic analysis shows closely spaced irregularly bounded organic remains, particularly coral. The internal wall structure of the bioclasts (fragments of biological origin such as shells) are poorly to moderately well preserved by a matrix of microcrystalline calcite (micritic matrix). Their internal chambers are later filled with calcite materials with slightly coarser size than micrite. Boundaries of the bioclasts are not well defined, hence the difficulty in recognizing them. Bioclasts are held together by micrite matrix. The mortar sample comprises of 90% bioclasts and 10% micritic matrix.

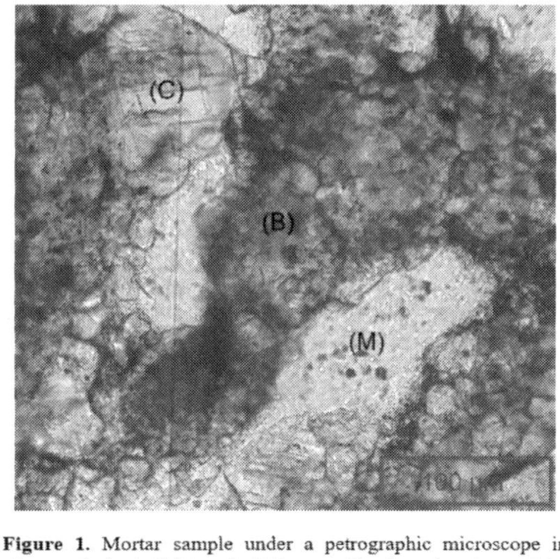

Figure 1. Mortar sample under a petrographic microscope in plane polarized light and 20X magnification showing calcite fragments (C), bioclasts (B), and micrite matrix (M).

3.2 Distribution of particles in the mortar

Sieve analysis was implemented to determine the particle size distribution of the historical mortar sample. Figure 2 shows the weight percentage of each fraction that was retained in the sieve. It is observed that 4.75

mm sieve retained the most weight followed by 1.18 mm, 0.6 mm, and 2.36 mm. Also, it was observed that some lime particles are still intact with the aggregates. The fraction that was retained in the pan (<0.075 mm) is correlated with the calcium carbonate binder [4].

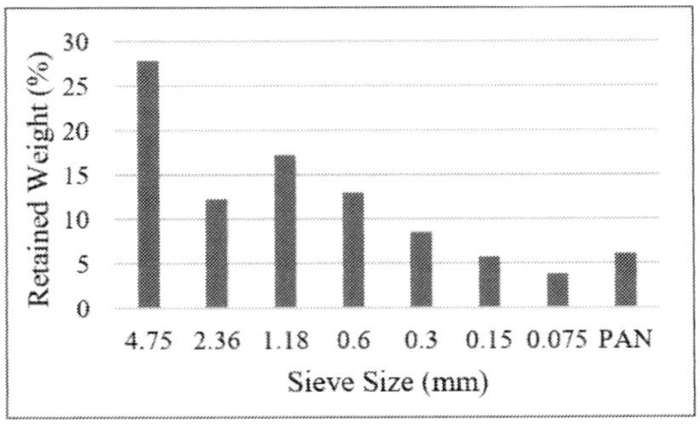

Figure 2. Weight distribution in percentage of each sieve fractions of the historical mortar sample.

It is shown in figure 3 the graph of the cumulative percentage which is also known as the percent finer (% finer) [4]. This data will signify if each sieve fractions were well represented pertaining to all sizes of the particle. The value for D_{10}, D_{30}, and D_{60} are 0.15 mm, 0.8 mm, and 2.9 mm respectively. The values acquired were gathered via extrapolation. These values designate to the diameter (D) of the grains that pertains to the 10%, 30%, and 60% of the sample passing through the sieve by means of weight. With these parameter values, the coefficient of curvature(C_c) was computed to be 1.47 mm and the coefficient of uniformity (C_u) was computed equal to 19.33 mm. With the value of C_u to be greater than 4 and the value of C_c is between 1 and 3, the historical mortar sample is considered to be well - graded and well represented over a wide range [5].

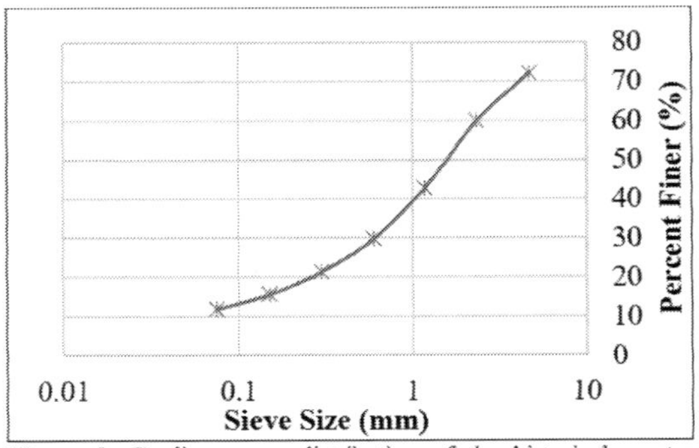

Figure 3. Grading curve distribution of the historical mortar sample.

3.3 Atomic absorption spectroscopy (AAS)

The sieve fractions of the historical mortar sample were digested with HCl to determine the distribution of calcium in sample. Taking into consideration that historical mortars are usually made from limestone ($CaCO_3$) which is basically the binder matrix of the historical mortar. Furthermore, the components that was digested by HCl solution would be considered as the binder, while the insoluble components would be the aggregates [4]. Table 1 represents the distribution of calcium content from the historical mortar sample. Looking at the Ca content of 4.75 mm, it is observed that it has a lower % Ca compared to calcite which is about 40 %Ca. This pertains that during the burning process of limestones, not all $CaCO_3$ would turn into CaO but instead, some Ca atoms would react to other components, which could be from the soil added during the process.

Table 1. Summary of the Atomic Absorption Spectroscopy (AAS) data for calcium.

Sieve Size (mm)	Mass of Mortar Sample (g)	Ave. % Composition of Ca
4.750	1.661	13.494
2.360	0.728	6.149
1.180	1.076	5.125
0.600	0.822	4.035
0.300	0.514	4.670
0.150	0.343	6.939
0.075	0.247	4.950
< 0.075 (pan)	0.372	3.506

3.4 Thermal analysis

As the temperature was increased to 900 °C, weight loss measurement of the historical mortar sample is established to acquire information about the sample's binder matrix. It is shown by the thermograph in figure 4 its continuous weight loss and its significant peaks between 740 °C and around 850 °C. This particular drop in the sample's mass pertains to a characteristic peak of decomposition of calcium carbonate, with its release of CO_2 [6]. The significant drop of sample's mass between 200 °C and around 590 °C is attributed to the evaporation of water which is chemically bounded to hydraulic compounds, and the decrease in weight above 590 °C is due to the loss of CO_2 [7]. The first derivative plot indicates that at weight loss was highest at 801.96 °C.

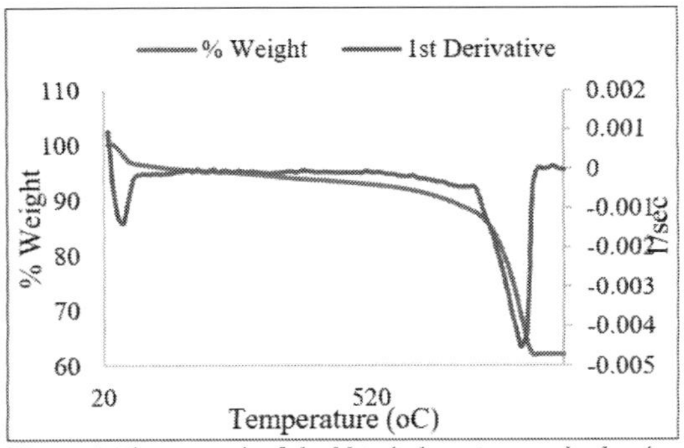

Figure 4. Thermograph of the historical mortar sample showing weight loss as temperature increases.

3.5 IR Spectroscopy characterization

Shown in figure 5 is the corresponding IR absorption peaks of the inorganic materials present in the historical mortar sample. Characteristic peaks of the binder matrix which is primarily composed of $CaCO_3$, the important absorption bands for C-O are the out-of-plane bending (υ2) at 874 cm^{-1}, in-plane bending (υ4) at 714 cm^{-1}, and 1425 cm^{-1} that corresponds to the asymmetric stretching (υ3) [8]. Furthermore, the broad peaks at 2511 cm^{-1}, 1155 cm^{-1}, and 1063 cm^{-1} corresponds to the C-O asymmetric stretching and is attributed to $CaCO_3$ respectively. The presence of sand which is essentially silicon dioxide, that would pertain to the silica found in soil by which the limestone was acquired, thus absorption bands for quartz is also present. Based on the Si-O stretching peaks at 667 cm^{-1}, and 1063 cm^{-1} respectively [4].

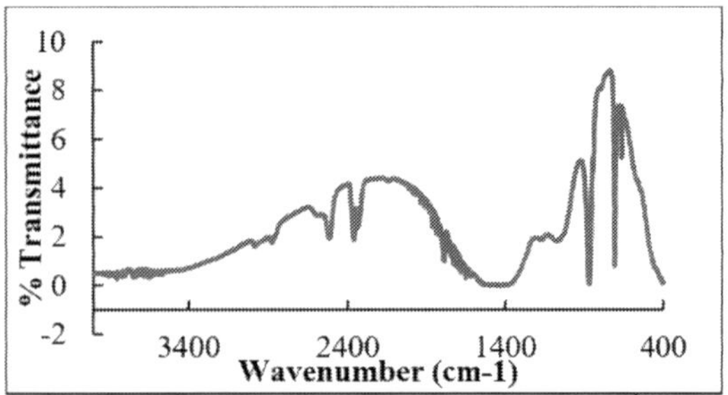

Figure 5. Mid-infrared region spectrum (4000 – 400 cm^{-1}) of the historical mortar sample.

4 Conclusion

This paper presents different analytical methods which can effectively characterize historical mortars used in the Spanish-colonial period in the Philippines and can be used as a reference information for conservation and renovation of cultural heritage structures during the Spanish-Colonial period. Petrographic analysis determined the presence and majority of bioclasts which is basically fragments of shells or corals. This implies the use of shells or corals in creating the historical mortar. The majority of micritic matrix within the historical mortar sample is due to the major component of historical mortars which is limestone. The sieve analysis expressed its well-graded particle distribution. Atomic absorption spectroscopy provided the calcium content present in each sieve fractions. The acid soluble constituent is attributed to the binder (limestone) component, while the insoluble components is considered to be the aggregates. The low %Ca (~13%) of the sample pertains to the interaction of Ca atoms to other constituents present in the soil were the limestone was gathered, which is basically added during the burning process of lime. Thermal analysis shows the weight loss of the historical mortar sample as the temperature increases. It was speculated that the non-linear drop between 740 °C and 850 °C corresponds to the decomposition temperature of calcium carbonate. And lastly, IR analysis provided basic mineralogical characteristics of the historical mortar sample, primarily of $CaCO_3$ and other constituents such as SiO_2.

5 References

1. J. Salceda, *Presentation in the 4th Membership Meeting of PTAA and 18th Lakbay Pilipinas.*
2. A. Estevao Candeias, P. Nogueira, J. Mirao, A. Santos Silva, R. Viega, M. Gil Casal, I. Ribeiro, A.I. Seruya, "Characterization of ancient mortars: present methodology and future perspectives," *Chair. of the Euro Res Coun Chem Com*
3. A. Moropoulou, A. Bakolas, "Investigation of the Technology of Historic Mortars," *J. Cul. Her.*, **1**, 45-58, [2000]
4. J.M.C. Cayme, A.N. Asor. "Characterization of historical lime mortar from the Spanish colonial period in the Philippines," *Con. Sci. Cul. Her.*, **1**, 33-57, [2016]
5. R.K.W. Holtz, "An introduction to geotechnical engineering," *Pren. hall, Inc.* [1981]
6. J.M.C. Cayme, A.N. Asor, Jr., M.K.A.T. Alano, E.T. Miranda, "Chemical characterization of historical brickwork of the church convent in Pagsanjan, Laguna." *KIMIKA*, **27**, 30-41 [2016]
7. P.J.P. Gleize, E.V. Motta, D.A. Silva, H.R. Roman. Char Hist Mort Sta Cat Bzl. *Cement and Concrete Composites*, **31**, 342-346, [2009]
8. P. Sathya, G. Velraj, S. Meyvel. Four Trans Inf Spec Stud Anc Sam Sal Reg Tam Ind. *Advanced Applied Science Research*, **3**, 776-779, [2012]

Petrographic and Thermal Characterization of Mortar from a Church Ruin in Cagsawa, Albay, Philippines

Nina Muriel A. Laplana[1], Eric T. Miranda[1], John Carlo A. Mangay[1], Angel S. Recto[2] and Jan-Michael C. Cayme[1]

[1]*School of Chemical, Biological, Materials Engineering and Sciences, Mapua University, Muralla St., Manila, Philippines, 1002*
[2]*College of Social Sciences and Philosophy, Bulacan State University, Malolos, Bulacan, Philippines, 3000*

Abstract. The Cagsawa Ruins is a crucial structure of the old church standing 38 km from Mt. Mayon in the province of Bicol. The ruins has been the result of the eruption of Mt. Mayon last February 1, 1814. The remains have become a part of the province's rich history and culture declaring it as one of the national treasures of the Philippines. Preservation is implemented to maintain the part that has remained from centuries ago. But apparently, the material used for preserving this important structure lacks compatibility with the original materials incorporated in the structure. Portland cement has been an alternative for lime mortars through which, in nature, are synthetic and are known to be the source of premature deterioration of soft, historic bricks. This cultural heritage is of importance to the history and must be preserved to its rightful material. By means of petrographic and thermal analysis it was proved that the aggregate from the mortar gathered from Cagsawa came from minerals and marine organisms that the decomposition of the elements confirmed that there are organic compounds present.

1 Introduction

For almost two centuries the Cagsawa Ruins has stood as a symbol of Bicol region's impressive landscape, rich history and the people's strength and resiliency to face and to rise from the ravages of Mother Nature. The Cagsawa Ruins are what remained of the old Cagsawa community when the near perfect cone-shaped Mayon Volcano erupted early in the morning of February 1, 1814. Accounts of the eruption said columns of rocks and stones shot high into the air killed fleeing residents and set houses on fire as the volcanic debris fell down to earth. Over time, the church building collapsed and today, only the blackened church tower remains standing which is now the centerpiece of the Cagsawa Ruins Park. The Cagsawa bell tower is one of the popular destinations in Albay. Historical structures bring character and certain charm to the neighborhood that people live in. Thus, the preservation and restoration plays a cultural role. Old buildings teach us about the history that happened before we were born and promotes the respect for those who lived in different times and societies [1]. It is important to understand the proper processes of restoration of these old structures using materials that compliments the original structure. Furthermore, this study will benefit other fields of discipline like architecture, history, and archeology.

A mortar is a material resulting from the mixture of sand grains, binder (lime) and water [2]. The properties and characteristic of the mortars mainly depend on the nature of the binder component. That is the reason for which, its evolution with time has been very much related to the development of artificial cementitious materials. Portland cement as a modern day alternative has properties that are incompatible with lime mortar [3]. Aside from its physical difference, Portland cement is harder, less flexible, and impermeable. These qualities lead to premature deterioration of soft, historic bricks. The use of historical methods of low temperature fired, lime mortars with existing mortar of a similar type is a much better treatment for historical building restoration.

This study aims to characterize an old lime mortar that was used in the construction of the historical church ruins in Cagsawa, Albay. This paper will try to investigate by petrographic analysis and thermal analysis, the composition of the lime mortar and the reactions that may occur in the samples undergoing temperature differences [4] that was used in building the ruins of Cagsawa to open an idea to present renovators on how to preserve these important structures.

The utilization of the Petrographic analysis leads us to assess the Physical characteristics of the lime mortar that was gained from the ruins. The chemical analysis of the mortars has also been carried out by using the Thermogravimetric Analyzer on where the samples are being heated up to 1000C in order to show response of the weight and the samples' decomposition in terms of increasing temperature.

This study mainly encompasses the characterization of mortar used in the ruins of Cagsawa in Barangay Busay, Albay which will strictly focus on the microscopic analysis of the mortar samples. The mortars mineralogical constitution is very important to understand the mechanical behavior [5]. There are four samples gathered from different areas in Cagsawa. The first sample is from the west wall of the bell tower, the second sample is from the north wall of the convent, the third and fourth sample is from the east and west wall of Casa Real.

This study is part of a research project approved by the Mapua University, Directed Research for Innovation and Value Enhancement Project (DRIVE) in 2016 Titled: Chemical Analysis of Historic Mortars from Different Spanish Colonial Fortifications in the Philippines.

2 Materials And Methods

Four samples were gathered from the ruins of Cagsawa. The samples were detached from broken concretes situated on the ground of the vicinity. The samples were named according to the place where it was obtained (see Table 1 for reference) and was stored in a zip lock bag. Fragments of these samples were subjected to thin section preparation and petrographic analysis. The importance and the limited resources of these ancient materials leads us to handle them carefully for experimentation. Each of the samples were sawn by a 10 inch saw together with an oil lubricant in order to avoid saw rusting. The oil used together with the saw forms a thick sludge that is used to glue the 20μm thickness of the sample in the glass slip. The four samples are examined with petrographic microscope under magnifications 20x and 4x.

Table 1. Name of the sample and the Origin where it was obtained.

Name	Origin
C-01	West wall
C-02	North wall
C-04	East wall
C-05	West wall

Two samples, C-02 and C-04, were subjected to thermal analysis. A STARe Default DB V11.00: METTLER has been used to determine the change of the sample as it is subjected to increasing temperature. A portion of the samples were powdered and weighed to 6.8 micrograms. The sample was then placed on the aluminum pan next to the reference pan, the balance was then tared; the system was purged with nitrogen gas as the sample pan and the reference pan were set to heat at a temperature range of 30°C (room temperature) to 1000°C, with a heat rate of 10 °C/minute under dynamic air with a purge rate of 20 mL/minute. The balance then sends the weight signal of the sample pan and the reference pan along with the total time elapsed to the computer. The abscissa (X-axis) is displayed as time or temperature and the ordinate (Y-axis) is displayed as weight (mg) or weight percent (%).

3 Results and Discussion

3.1 Petrographic Analysis

Thin sections determines the mineralogy, the quality and the composition of the original lime mortars; microscopic cracks formed in elements and the causes of their occurrence (crystallization under pressure, tensions between mortar and rock), porosity of materials, chemical or biological attack. [8]

Petrographic analysis show homogenous, cohesive binders displaying a strong binder aggregate bond and an absence of over burned and under-burned lime particles. The samples show uniformity as it depicts clastic particles with a conglomerate of rock forming minerals. Each samples are greatly compose of lithic fragments which are a derivation from sedimentary, igneous and metamorphic rocks. They are particles that are eroded down into grains or into sand size. All of the samples are greatly surrounded by micrite and goethite matrix which are constituents of limestone and iron hydroxide minerals. Small percentages of silicate minerals are visible in each sample. Figure 1 shows the 20x magnification view under plane polarized light image of the sample obtained from the west wall of the bell tower. Plagioclase and pyroxene minerals are in minimum percentages as represented by the white particles.

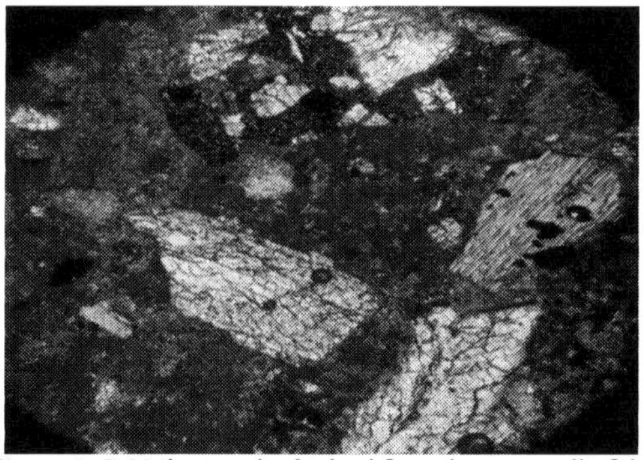

Figure 1. C-01 the sample obtained from the west wall of the bell tower in 20x magnification under PPL (Plane polarized light)

Both of these minerals are silicates, which are made up of silicon oxides and are commonly found on the earth's crust. They are also derived from igneous, metamorphic and sedimentary rocks. The rock particles are all identified as volcanic types through which we can easily tell as it is situated near the Mayon volcano. Fragments of rocks and binder/aggregates that came with the mortar are of easy access.

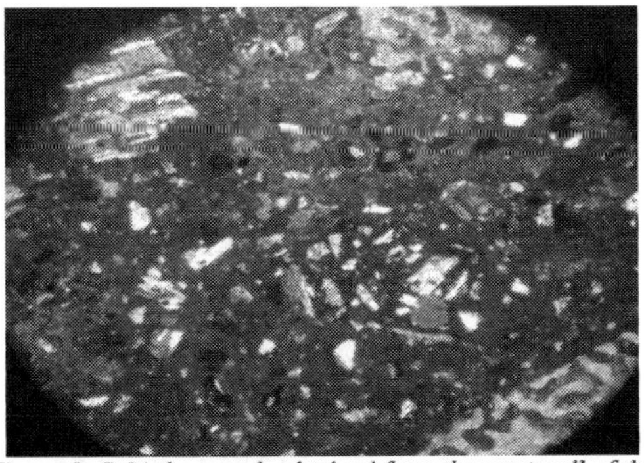

Figure 2. C-01 the sample obtained from the west wall of the bell tower in 4x magnification under PPL (Plane polarized light)

The 4x magnification view in Figure 2 of sample C-01 shows fragments of fossils of living organisms in the marine life. Seashells makes up a bioclast of 40% in the samples gained that is visibly represented by Figures 1-5. These external skeleton of some marine organisms are known to be rich in calcium oxide content which is an element of decomposition of limestone after being heated at a certain degree.

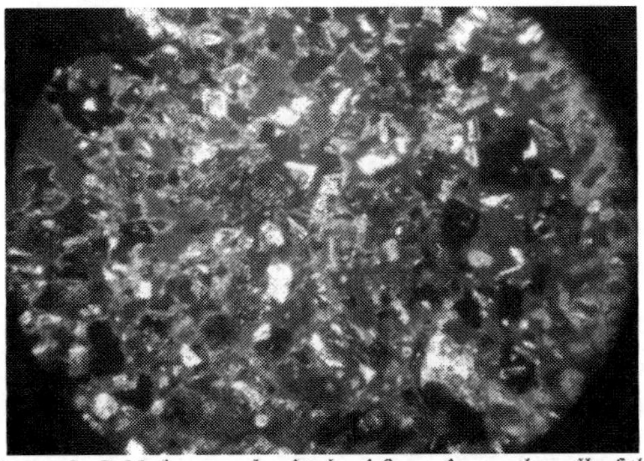

Figure 3. C-02 the sample obtained from the north wall of the bell tower in 4x magnification under PPL (Plane polarized light)

Figure 4. C-04 the sample obtained from the east wall of the Casa Real in 4x magnification under PPL (Plane polarized light)

Figure 5. C-05 the sample obtained from the west wall of the Casa Real in 4x magnification under PPL (Plane polarized light)

Crystal twinning also shows on the cross polar image of our C-01 sample on Figure 6. The twinning of crystals occurs when it is subjected to stress or pressure/temperature different from the original process of its formation. Plagioclase is the mineral crystal that forms twinning in the sample. The lattice points of the crystal is formed and shared with the lattice points of another same crystal and there it forms twins of crystals. The said twinning can also tell that the crystals together with other organic fragments and carbon oxide elements are subjected to high temperature and pressure as the mortar is being done.

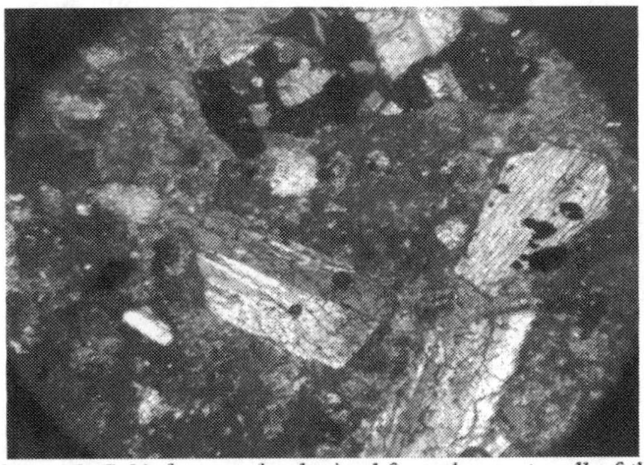

Figure 6. C-01 the sample obtained from the west wall of the bell tower in 20x magnification under XPL (Cross polarized light)

3.2 Thermal Analysis

Thermogravimetric analysis shows the nature of the aggregates and other aspects of the mortar when subjected to increasing temperature. This method reveals thermal transformations, which include dehydration, dehydroxylation, oxidation and decomposition [9]. Figure 7 shows that the sample, C-02, decomposes minerals at 680 C. Partial oxidation of the sample is visible at 40 C which resulted to a weight increase of the sample [6]. As temperature increases, the weight of the sample decreases. The decrease in weight that ranges between 100 C to 600 C is a result of the evaporation of water as well as the combustion of organic matter. The First derivative curve represented by the red line shows a rapid decrease in weight of the sample at 670 C visibly indicates a loss of minerals. This rapid decrease can also be the result of the decomposition of the Calcium Carbonate to Calcium Oxide and Carbon Dioxide.

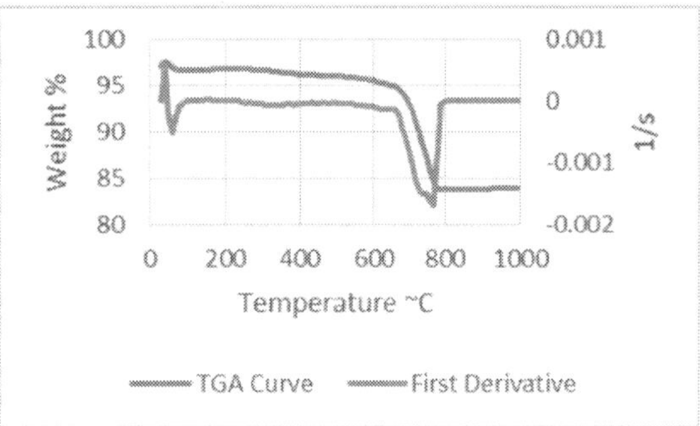

Figure 7. Thermogravimetric Curve of the sample C-02 and its First derivative curve represented by the red line.

Same results also occurred to the sample acquired from the east wall of the bell tower. The First derivative of the curve in Figure 8 also shows an exothermic peak at the initial point. The TGA shows a slight increase of the curve from the initial point up to 600 C which also indicates a slight increase in weight of the sample. A constant decrease in weight of the curve occurred from 600 C up to 690 C indicates a loss of water or moisture incorporated in the sample. The rapid decrease in weight of the sample at 690 C can greatly be attributed by the decomposition of minerals and CaCO3 from the lime binder.

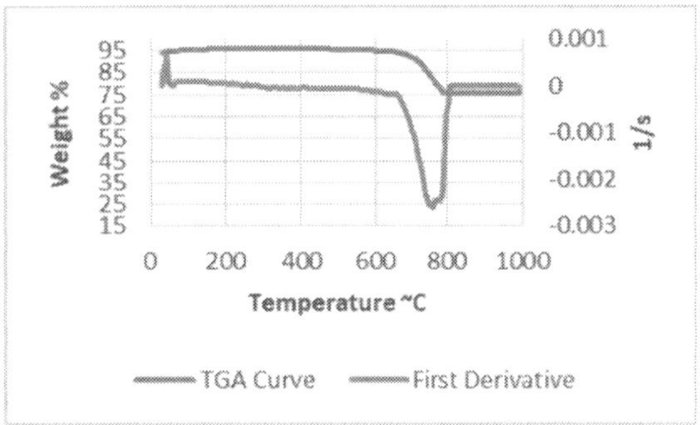

Figure 8. Thermogravimetric curve of the sample C-04 and its first derivative curve represented by the red line.

4 Conclusion

The uniformity and the compactness of the binder/aggregate has been evidenced by the petrographic microscope. There are no over-burnt or under-burnt particles. The theoretical reason for this is that the burning of the mortar is strictly observed and being added with pieces of wood from time to time [7].

The twinning of the crystals tell that the process for making the mortar is defined by adding a certain amount of pressure and a certain degree of temperature.

The reactions of the two specimen that have undergone thermogravimetric analysis shows similar reactions under temperature difference. The melting peaks that shows the rapid decrease of weight of the samples are both an indication that reactive minerals have decomposed.

The data gathered from the petrographic analysis used in this study has determined the main composition of the mortar. It is evidenced that the mortar is composed of external skeleton of some marine organisms that makes up the aggregate of the material from which can be concluded that the materials used came from local sources. It also emphasizes the lime that makes up the binder of the mortar.

The results gained from the two techniques makes a crucial amount of information and has met similar results as displayed by the decomposition of the elements in the mortar seen in the TGA graphs is a confirmation that organic compounds are present.

5 References

1. R.Archivotli, Imp. of Save. H. MONT (2014)
2. E.S Goins, *Standard Practice for Determining the Components of Historic Cementitious Materials*, **6** (2002)
3. L. Strenchock, Res. On Rest. Of HTE Struc. (2009)
4. J.I. Alvarez, I. Navarro, P.J. Garcia Casado. Therm., Min. and Chem. Stud. Of the Mortars used in the Cathedral of Pamplona(Spain)
5. P.C. Raposo, J.A.F.O. Correia, D. Sousa, M.E. Salavessa, C. Reis, C. Oliveira, Abilio De Jesus PETRA. Char. Part. Wall Mortars of a 19th Cent. Bldg (2017)
6. Jan-Michael C. Cayme, Aniano N. Asor, Jr., Maveen Kim Alexis T. Alano and Eric T. Miranda. "Chemical Characterization of Historical Brickwork of the Church Convento in Pagsanjan, Laguna" (2016)
7. Regalado Trota Jose, *Zero In* (2003)
8. N. Cobirzan, A.-A. Balog, Characterzation of Mortars Compatibility using Microcopical and XRD Analysis (2013)
9. A. Moropoulou, A. Bakolas, K.Bisbikou. Characterization of Ancient Byzantine and later Historic Mortars by Thermal and X-ray Diffraction Techniques (1995).

MATEC Web of Conferences **213**, 02003 (2018) https://doi.org/10.1051/matecconf/201821302003
ACMME 2018

Study on the Estimation of Rock Rheological Parameters under Multi-level Loading and Unloading Conditions

Han Bing, Fu Qiang

China National Institute of Standardization, 100191, Beijing, China

Abstract. Determining reasonable rheological constitutive models based on laboratory rock creep test data and estimating rheological parameters are important means of studying rheological mechanical property of rocks. Previous rock uniaxial or triaxial compression creep tests mostly adopted the method of multi-level loading. In order to study the influence of unloading stress on the estimation of rock rheological parameters, rock triaxial creep tester RLW-2000 was used in this paper to carry out triaxial compression creep tests for granites in a given area with two different stress paths of multi-level loading and stepwise unloading. Study results indicate that rock rheological parameters obtained vary with different stress paths. The instantaneous elastic modulus and viscosity coefficient obtained through loading show a stepwise increase, while the viscoelastic modulus shows a progressive decrease.

1 Introduction

Rheological property is an important mechanical property of rocks. It is closely related to long-term stability of geotechnical engineering. Determining reasonable rheological constitutive models based on laboratory rock creep test data and estimating rheological parameters are important means of studying rheological mechanical property of rocks. Lots of references have mentioned the study on uniaxial and triaxial compression creep tests of rocks under loading conditions in recent years[1-6], but there are few studies on rock creep tests and rheological mechanical property under unloading conditions. Practice indicates that engineering rock masses are inherently different under two different stress conditions of loading and unloading. In many rock mass projects including underground caverns, rock slopes and rock foundations, the load or surrounding rock stress imposed on the rock mass is generally released in a stepwise manner as the construction progresses. It is always in the stress environment of progressive unloading. The rock creep under such stepwise unloading conditions is an important physical and mechanical property and engineering phenomenon of rock masses. If corresponding rock mechanical parameters are obtained based on traditional loading mechanical test results only, the data are often greatly different from actual observation results. Therefore, carrying out rock creep tests under unloading conditions and studying estimates of corresponding rheological mechanical parameters have great theoretical value and actual engineering significance.

In this paper, with surrounding rocks of underground storage to be built in a coastal area as the object of study, laboratory triaxial compression creep tests were respectively carried out for granite specimens in the storage with two different stress paths of multi-level loading and stepwise unloading. Test results were used to identify the rheological constitutive model, thus obtaining the change law of rheological mechanical parameters under two stress conditions of loading and unloading.

2 Rock Triaxial Creep Test

Rock triaxial creep test is an important means to study rheological mechanical property of rocks. Test results are the base of rheological constitutive model study. Actual test data are of great importance to the inversion of known rheological model parameters and identification of unknown rheological models. Considering that engineering rock masses are always under three-dimensional stress conditions, rock triaxial creep tests were carried out in this paper for the purpose of fully learning about rheological mechanical property of surrounding rocks of the underground storage.

2.1 Test equipment

Rock triaxial rheological tester RLW-2000 was used to carry out laboratory creep tests. This tester consists of an axial loading system, a confining pressure loading system, a servo system, a control system, a data acquisition and automatic drawing system and so on. Fig. 1 shows the tester.

Control parts of axial and confining pressure loading systems adopt EDC which allows force control or deformation control, also supports smooth switching between the control methods during the test. The pressure stabilizing system uses an AC servo motor for automatic pressure stabilizing. The deformation measuring device uses displacement extensometers from the American Teratech Corporation, including two sets: axial and radial extensometers, with the relative error of ±0.5%.

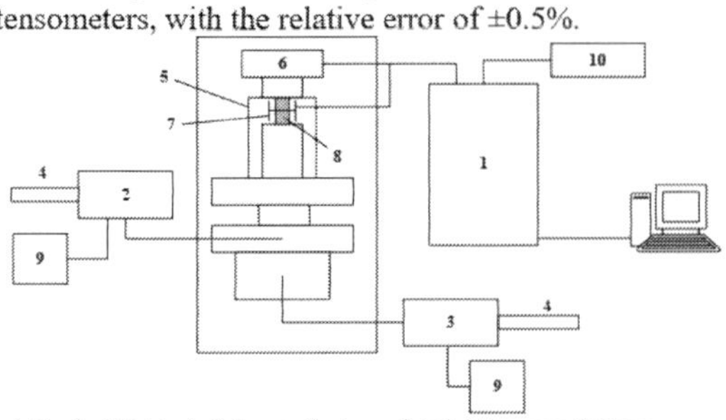

1-Control Cabinet, 2-Power System of Axial Pressure, 3-Power System of Confining Pressure, 4-Ball Screw, 5-Pressure Cell, 6-Pressure Sensor, 7-Deformation Sensor, 8-Rock Core, 9-Servo Motor, 10-Power Supply

Figure 1. RLW-2000 triaxial rheological test system of rock.

2.2 Test conditions and plans

Rock cores used in the test were all taken from the project site. Rocks in the storage are mainly medium- and coarse-grained syengranites of the overbridge unit in the early cretaceus period of the mesozoic erathem, with good appearance integrity and uniformity. Three typical places were determined on site for drilling and sampling analysis. Rock cores were made into cylindrical standard specimens in a unified way, with a diameter of 55×110 mm. The method of triaxial creep test was adopted for temperature-controlled long-lasting test. Well-prepared rock cores were sealed with oil-proof corrosion-resistant shrinkable tubes and placed in the self-balancing pressure chamber of the rock axial creep tester. With the central position well adjusted, the axis of the specimen coincided with the loading center line of the tester, avoiding nonhomogeneity of rock loading resulting from eccentric compression. The test was carried out in a special creep laboratory under constant temperature and humidity conditions. Temperature in the laboratory was strictly controlled at 25±0.1°C during the test. The computer collected the data of load, deformation and time automatically during the test, and displayed the axial strain-time relation curve in real time.

In order to fully consider the influence of different stress paths on rock rheological property and mechanical parameters, two plans were designed for this test: multi-level loading and stepwise unloading.

Plan (1): The confining pressure keeps constant, while the axial pressure uses the method of multi-level loading, namely, based on the instantaneous breaking strength obtained from conventional triaxial compression test, divide the maximum load to be imposed into several levels, and then impose the load on the same specimen gradually from the lowest level to the highest level. The imposing duration of the load at each level should depend on the strain rate or stress rate variation of the sample. Conduct low-stress pre-compression for the specimen first, and observe the displacement continuously. Generally, the pre-compression period is controlled within about 24 hours. Then impose the preset confining pressure on the specimen through the hydraulic system at a loading rate of 0.05~0.02MPa/s, to make the specimen under hydrostatic pressure. When the confining pressure reaches the set value and deformation becomes stable, impose axial load at a rate of 500N/s. When the load reaches stress of the first level, keep the axial stress unchanged, observe and record the relationship between axial strain and time. After the creep deformation of the specimen becomes stable, impose the axial load of the next level, and maintain the stress of this level imposed on the specimen constantly. Repeat the above steps, and stop the test until the rheodestruction of the specimen under the preset stress of the last level.

Plan (2): The confining pressure keeps constant, and the axial pressure uses the method of stepwise unloading. Divide the maximum load to be imposed into several levels, and then decrease the axial load on the same specimen level by level from high to low. In order to avoid that excessive primary load causes the specimen to be damaged instantaneously or enter the accelerated creep stage ahead of time, set the stress of the first level to be 80% of instantaneous breaking strength, keep the load of the level unchanged, observe and record the relationship between axial strain and time, and conduct the first level of unloading for the level of stress after the creep deformation of the specimen becomes stable. The load decrement is the same with the load increment in the loading test. Stabilize the level of stress and keep it unchanged after the preset load of the next level is reached, until the start of the next level of unloading. The determination of duration for each level of load is as same as that in Plan (1). As granite specimens used in the test are hard brittle rocks whose rheological characteristics are not obvious under low stress, the test should be stopped when the axial load decreases to 40MPa.

2.3 Analysis of test results

Fig. 2 and 3 respectively show triaxial creep test curves of specimens 3-26 taken from Drill No. 3 under two stress path conditions of stepwise unloading and multi-level loading. Unloading test was carried out first. It can be seen from Fig. 2 that reversible deformation occurred rapidly on specimens at the moment of unloading, indicating an instantaneous elastic recovery of rocks. Hysteretic rebound deformation, i.e. elastic aftereffect, happened over the time. The specimen had instantaneous strain at the moment when each level of stress was released. The strain capacity increased with the decrease of stress level. Under most levels of stress, instantaneous strain accounted for the main part of overall deformation. For the loading test in Fig. 3, the deformation-time curve of the specimen shows that the creep value under low stress was relatively low; deformation became stable after a short initial deformation; the creep deformation rate of rocks decreased gradually with the time under relatively high stress; finally deformation tended to be restrained and entered the stage of steady flow; under the stress of the last level, rock deformation was aggravated sharply, showing a tendency of obvious accelerated creep. Cracks appeared outside the specimen and expanded quickly. Rapid evolution and accumulation of damage finally resulted in rock rupture.

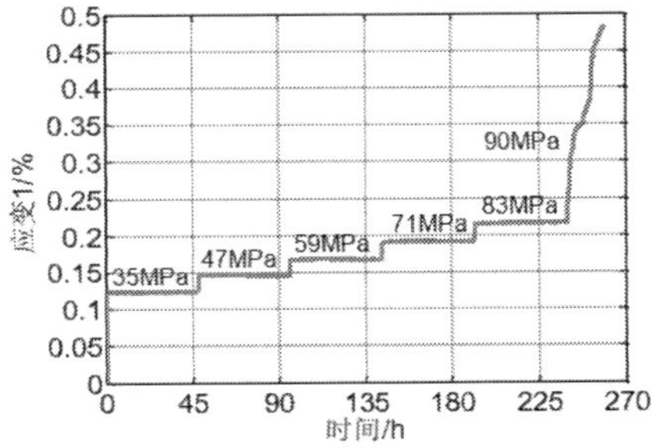

Figure 2. Triaxial creep curve of rock under step unload.

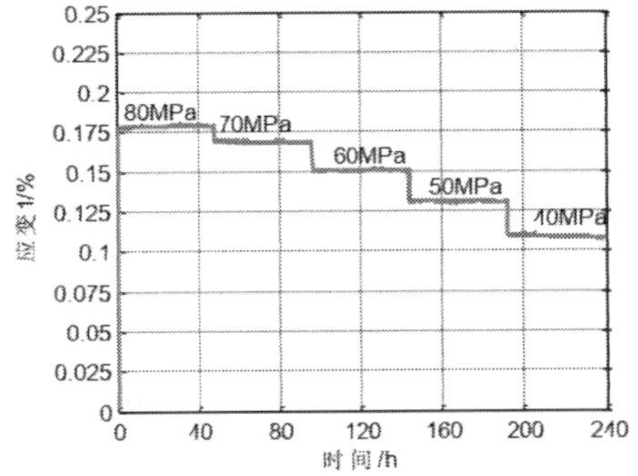

Figure 3. Triaxial creep curve of rock under step load.

3 Model Identification and Parameter Estimation

To carry out rheological study on rocks, it is important to select proper rheological models according to laboratory rock creep test data and determine corresponding rheological parameters through parameter inversion or curve fitting. Rheological constitutive models of rocks often fall into three types: empirical formula, combined model and integral model, among which combined model consisting of elastic, viscous and plastic basic elements are most widely used.

3.1. Rheological model identification

Rheological characteristics of high-strength rock materials like granites are not obvious under low stress. According to Fig. 2 and 3, strain tended to be restrained gradually with the time under most levels of stress; both the elastic strain that recovered instantaneously and elastic aftereffect happened at the moment of unloading, with rheological property conforming to the feature reflected by the generalized Kelvin model. Therefore, the three-parameter generalized Kelvin model was used to describe the mechanical property, as shown in Fig. 4, with the constitutive equation as below:

$$\varepsilon = \frac{E_0+E_1}{E_0E_1}\sigma - \frac{\eta_1}{E_0}\dot{\varepsilon} + \frac{\eta_1}{E_0E_1}\dot{\sigma} \tag{1}$$

Wherein, E_0 and E_1 respectively refer to instantaneous elastic modulus and viscoelastic modulus while η_1 is viscosity coefficient.

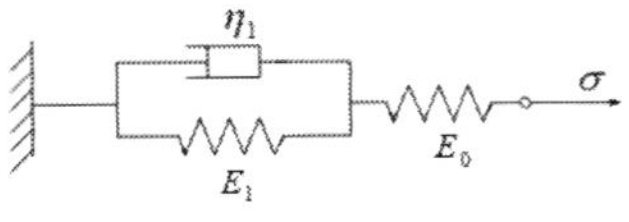

Figure 4. H-K rheological model.

3.2 Methods of parameter estimation

Presently, methods used to determine rock rheological parameters according to laboratory test data and curves mainly include regression analysis[7], optimized separation[8], least square method[9-10], curve decomposition[11-12] and so on. In this paper, the method of optimized separation for viscoelastic parameters put forward in Reference[8] was used to calculate rheological parameter. Firstly, designate the design variable and objective function. Take the parameter to be inverted as the design variable. Calculate the strain residual sum of squares based on the sum of n test data ε from the creep test data, and take the minimum value as the objective function for optimized separation of parameters. So, the design variable is

$$X = (E_0, E_1, \eta_1) = (X_1, X_2, X_3) \tag{2}$$

The objective function is

$$f(X) = \sum_{i=1}^{n}[\varepsilon(t_i)_c - \varepsilon(t_i)_m]^2 \tag{3}$$

Wherein, $\varepsilon(t_i)_c$ and $\varepsilon(t_i)_m$ respectively refer to the strain value obtained through calculation and testing at the time of t_i.

For the generalized Kelvin model, the objective function is

$$f(x) = \sum_{i=1}^{n}[\frac{\sigma}{E_1} + \frac{\sigma}{E_2}(1 - \exp(-\frac{E_2}{\eta_1}t_i)) - \varepsilon(t_i)_m]^2 \tag{4}$$

In order to simplify the above optimization problem, possible upper and lower limit estimates of the design variable can be provided according to analogies of engineering experience, thus to establish the following constraint condition

$$a_i \le X_i \le b_i \qquad (i = 1,2,3,4) \tag{5}$$

Wherein, X_i is the No. i design variable; b_i and a_i are the upper limit and lower limit of X_i respectively. With the mathematical model established based on Formula (2), (4) and (5), the method of multi-constraint optimization can be used to approach the optimal point X^l =(E_1, E_2, η_1) step by step and solve the minimization of the objective function. In this case, X^l is the rock characteristic parameter vector to be solved. According to the above method, determine rheological parameters as shown in Table 1:

Table 1. Rheological parameters.

Test plans	Sress/ MPa	Rheological parameters		
		E_0(MPa)	E_1(MPa)	η_1(Mpa))
loading	35	4.655×10^4	8.479×10^5	1.033×10^7
	47	4.708×10^4	6.713×10^5	3.812×10^7
	59	4.792×10^4	5.256×10^5	4.987×10^7
	71	5.381×10^4	3.970×10^5	5.663×10^7
	83	5.944×10^4	2.142×10^5	7.012×10^7
	mean	4.697×10^4	5.312×10^5	4.501×10^7
unloadin g	80	5.454×10^4	3.097×10^5	5.663×10^7
	70	4.603×10^4	4.784×10^5	3.214×10^7
	60	3.866×10^4	5.930×10^5	2.547×10^7
	50	3.727×10^4	7.415×10^5	1.189×10^7
	40	3.649×10^4	8.293×10^5	0.680×10^7
	mean	3.840×10^4	5.904×10^5	2.659×10^7

3.3 Analysis of test results

It can be seen from Table 1 that for rheological parameters estimated based on the multi-level loading creep curve, the obtained instantaneous elastic modulus E_0 and viscosity coefficient η_1 show a tendency of progressive increase with the increase of stress, while the viscoelastic modulus E_1 decreases gradually; for rheological parameters obtained through the stepwise unloading creep curve, the instantaneous elastic modulus E_0 and viscosity coefficient η_1 show a tendency of progressive decrease with the decrease of stress, while the viscoelastic modulus E_1 increases gradually. The parameter change law is exactly the opposite of that of parameters under loading conditions. According to the comparison of the means of all rheological parameters obtained under two different stress paths, the parameter E_0 and η_1 obtained under loading conditions are 81.75% and 59.08% of relevant parameters under unloading conditions, while for the parameter E_1, the mean under unloading conditions is 89.97% of that under loading conditions. This shows that the two different stress paths of multi-level loading and stepwise unloading have a certain influence on rock rheological mechanical property and parameter estimation. In previous rock uniaxial or triaxial creep tests, rheological parameters were mostly estimated based on the multi-level loading creep curve. However, for rock mass projects featured by excavation, the rock stress and strain redistribution law obtained from numerical calculation based on such rheological mechanical parameters cannot reflect actual conditions of the project truly and completely.

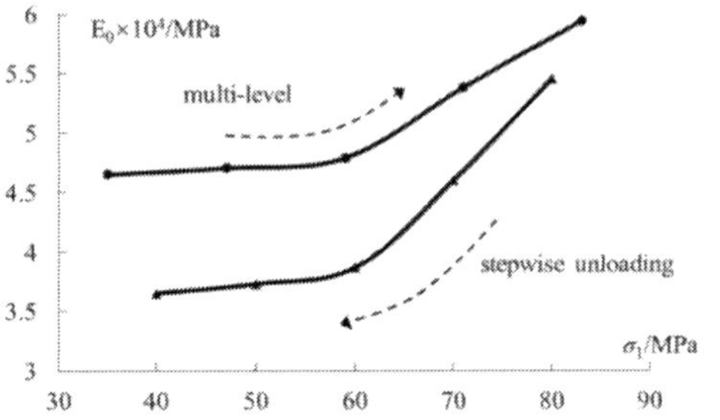

Figure 5. The relation between the elastic modulus E0 and stress paths.

Figure 6. The relation between the viscoelastic modulus E1 and

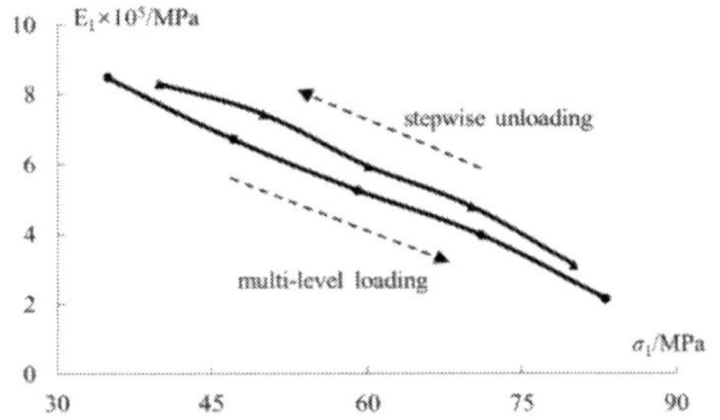

stress paths.

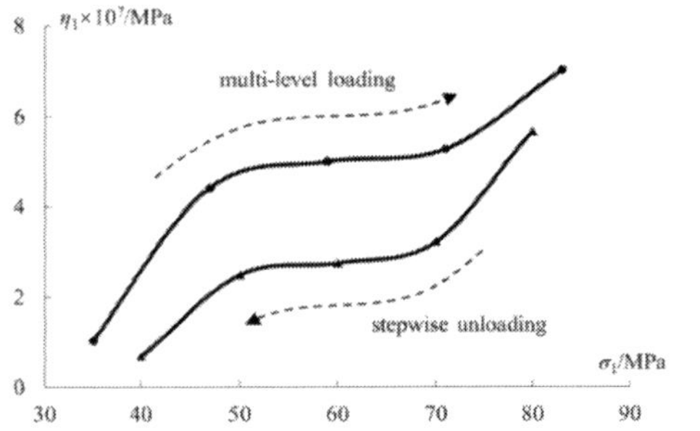

Figure 7. The relation between the viscous coefficient η 1 and stress paths.

4 Conclusions

Rock rheological constitutive model identification and parameter estimation are important issues in rock mechanical theory and engineering practice. In this paper, rock mechanical experimental study was carried out for granites in a certain area under two different stress path conditions of multi-level loading and stepwise unloading based on laboratory rock triaxial compression creep tests. On the basis of the creep curve in two stress states, the method of optimized separation was used to estimate rock rheological mechanical parameters, and the following conclusions were drawn through the analysis of test and calculation results:

(1) According to Fig. 5 and 7, Rheological parameters E_0 and η_1, obtained based on the multi-level loading creep curve, show a tendency of progressive increase with the increase of stress, while the viscoelastic modulus E_1 decreases gradually, as shown in Fig. 6; parameters E_0 and η_1 obtained through the unloading creep curve show a tendency of progressive decrease with the decrease of stress, while the viscoelastic modulus E_1 increases gradually.

(2) The comparison of means shows a great difference in rheological parameters obtained under loading and unloading conditions. Therefore, for the long-term stability analysis of specific rock mass excavation projects, if the numeric calculation is carried out based on rock rheological parameters obtained through loading tests, it can hardly reflect the change law of rocks in actual unloading environment. Hence, unloading creep tests should be carried out to analyze rheological property of rocks, so as to obtain corresponding mechanical parameters.

References

1. G. Liu and Y. S. Zhou: Seismology and Geology. Rheology of Felsic Rocks and Relative Influence Factors, Vol. 34 No.2, (2012), p. 365-383
2. J. Sun: Rock rheological mechanics and its advance in engineering applications. Chinese Journal of Rock Mechanics and Engineering, Vol. 26 No.6 (2007), p.1081-1115
3. W. H. Gao, Q. N. Chen, and Z. Y. Huang: Study on the creep damage constitutive model of soft rocks considering rheological softening and intelligent identification of the parameters. China Civil Engineering Journal, Vol. 45 No.2 (2012), p.104-110
4. G. J. Wang, L. Zhang, and Y. W. Zhang: Experimental investigation on the creep–damage-rupture characteristics of salt rock. Engineering Mechanics, Vol. 30 No.4 (2013), p.288-293
5. Z. W. Yang, A. B. Jin, and Y. T. Gao: Nonstationary Nishihara model in the particle flow code. Chinese Journal of Engineering, Vol. 37 No.7 (2015), p.831-838
6. T. Xu, C. A. Tang, and J. Zhao: Modelling the time-dependent rheological behaviour of heterogeneous brittle rocks. Geophysical Journal International, Vol. 189 No.3 (2012), p. 1781-1796
7. C. Z. Pu, P. Cao, C. Y. Zhang, and Q. C. Zhang: Variable Parameters Nonlinear Creep Damage Model of Rock with Consideration of Aging, Engineering Mechanics. Vol. 34 No.6 (2017), p. 17-27
8. Z.M. Shi and L. Zhang: Study on Rheological Test of Jinping Greenschist Under Step Load. Journal of Tongji University (Natural Science). Vol. 39 No.3 (2011), p. 320-326
9. M. F. Wang, B. Hu, H. F. Jiang, G. J. Ou, and L. Gan: Experiment and model investigation on shear rheological mechanical properties of granite. Journal of Central South University (Science and Technology). Vol. 45 No.9 (2014), p. 3111-3120
10. Y. Y. Huang and H. Zheng: Preliminary study of equivalent rheological model for jointed rock. Rock and Soil Mechanics, Vol. 32 No.12 (2011), p. 3566-3570
11. X. G. Wang, B. Hu, B. Q. Lian, J. L. Yan, Z. J. Xu, and Z. H. Zhao: Modified nonlinear viscoelastic-plastic rheological model and parameter identification of shear rheological model for granite. Chinese Journal of Geotechnical Engineering, Vol. 36 No.5 (2014), p. 916-921
12. S. G. Zhang, W. B. Liu, and Y. T. Wang: Study on Non-stationary Creep and Parameter Determination of Granite. Journal of Disaster Prevention and Mitigation Engineering, Vol. 37 No.3 (2017), p. 435-441

Synthesis and Characterization of Polysulfone (PSU)/Philippine Halloysite (PH-HAL) Nanostructured Membrane via Electrospinning

Ruth R. Aquino[1], Marvin S. Tolentino[1], Niel Karl G. Arcamo[1], John Patrick N. Gara[1], and Blessie A. Basilia[1,2]

[1]*Mapúa University, School of Chemical, Biological, and Materials Engineering and Sciences, 658 Muralla St., Intramuros Manila, 1002 Philippines*
[2]*Industrial Technology Development Institute, Department of Science and Technology, Bicutan, Taguig City, Metro Manila, 1631 Philippines*

Abstract. Membrane technology is widely used in many separation processes because of its multi-disciplinary characteristics. One of the techniques that is used in the fabrication of membranes is the electrospinning process which can create nanofibers from a very wide range of polymeric materials. In this study, electrospun nanostructured fibrous composite membranes of polysulfone (PSU), commercial halloysite (COM-HAL), and Philippine halloysite (PH-HAL) were synthesized. The concentrations of COM-HAL and PH-HAL were both varied from 0.5%, 1%, and 2%. The FTIR results showed that there were changes in the intensity of the PSU-IR spectra which confirmed the presence of COM-HAL and PH-HAL in the synthesized membranes. The SEM revealed that nanofibers can be successfully produced by the addition of LiCl salt in PSU with varying HAL concentrations. Also, it was observed that the addition of HAL with varying concentrations have no significant effect on wettability due to the strong hydrophobic character of the PSU membrane. Moreover, it was found from the analysis of mechanical properties that the tensile strength of the membranes weakened by the addition of HAL due to its weak interaction with PSU.

1 Introduction

Membrane technology is an emerging technology which can be used in many separation processes. Numerous membrane processes emerged in which applications are based on different separation principles and mechanisms. Various techniques and fabrication methods have been explored to synthesize membranes, and one of those techniques is the electrospinning process.

Electrospinning is a versatile process that can create nanofibers from a very wide range of polymeric materials. The practicality of electrospinning has been greatly improved with recent advances in mass production scalability, leading to higher production rate and lower cost materials. Moreover, the addition of salt concentration to polymer solution improves the nanofiber formation of the membrane [1]. One of the widely-used polymer materials for electrospinning is polysulfone (PSU). Since it exhibits excellent thermal, mechanical, chemical stability, and low cost, PSU is widely used in separation processes such as water and wastewater treatment, chemical, metallurgical, and bioseparation area. Previous research studied the influence of addition of nanoclay such as halloysite (HAL) on PSU nanocomposite to serve as excellent support for nanoparticles due to their unique geometry and properties. However, most of the work done on the addition of clay to PSU nanocomposite membranes had been prepared using dispersion methods.

This study focused on the synthesis and characterization of a nanostructured PSU/HAL membranes via electrospinning process. Specifically, this study aims to: (1) determine the effect of LiCl on the surface morphology of the nanostructured membrane; and (2) compare the following mixtures of electrospun membranes: pure PSU, PSU/LiCl, PSU/LiCl with commercial halloysite (PSU/LiCL/COM-HAL), and PSU/LiCl with Philippine halloysite (PSU/LiCl/PH-HAL) using different characterization techniques; namely, (a) effect of HAL addition on the chemical composition of the membrane by Fourier Transform Infrared (FTIR) analysis, (b) surface morphology through Scanning Electron Microscopy (SEM), (c) wettability of the surface with the use of contact angle goniometer, and (c) mechanical properties using the Universal Testing Machine.

Considering the low cost and availability of PSU and HAL, this study would serve as guide on the proper and efficient way of fabricating PSU incorporated with HAL. Also, this undertaking may serve as a gateway for further improvements and use of PSU/LiCl/HAL nanofiber membrane via electrospinning into various areas of applications.

2 Methodology

PSF, with an average molecular weight of 35,000, and COM-HAL were both obtained from Sigma-Aldrich. Lithium chloride (LiCl) salt and anhydrous dimethylacetamide 99.8% (DMAc) were obtained from the Mapúa University Chemical Supplies. PH-HAL was obtained from the Department of Science and Technology of the Philippines. The electrospinning procedure was partly based on methodology of Chang and Lin (2009) [1]. PSU mixtures were prepared at different compositions; specifically, pure PSU, PSU/LiCl, PSU/LiCl with varying concentrations of COM-HAL (0.5%, 1% and 2%), and PSU/LiCl with varying concentrations of PH-HAL (0.5%, 1% and 2%). 40 g of each polymer solution was prepared and stirred at 60 °C for 6 h. Initially, the solvent was prepared in a media bottle and preheated by submerging it in a water bath at 60 °C for 15 mins prior to dissolving the solutes. The flow rate of the solution and the voltage was set to 1.0 mL h-1 and 25 kV, respectively. The tip-to-collector distance was set to 15 cm. The equipment used to detect the chemical components and functional groups of the membranes was PerkinElmer Spectrum 100 FT-IR Spectrometer. The electrospun membranes were subjected to SEC Mini-SEM SNE-3200M micrograph analysis with 10 kV accelerating voltage for the comparison of surface morphologies. Moreover, 25 fiber diameters were measured using ImageJ ver. 1.50i software to calculate the average fiber diameter. Also, contact angle goniometer was used to determine the wettability of the membranes. The mechanical properties of the membranes were determined using Instron 3225 Single Column UTM.

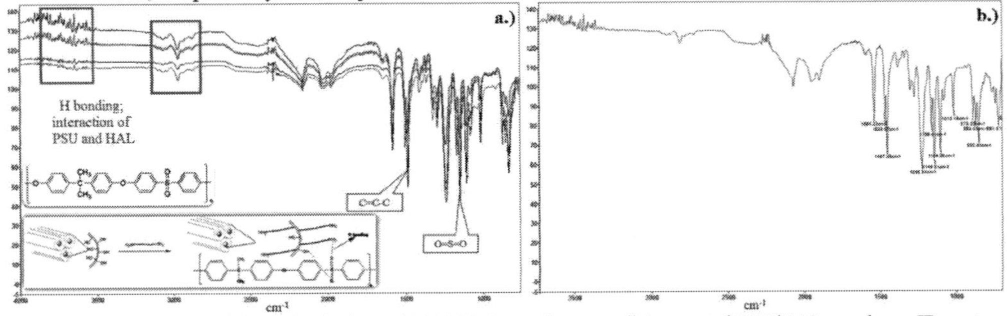

Figure 1.a.) IR spectrum of thesynthesized PSU/COM-HAL membranesandb.) expected PSU/HAL membrane IRspectrum.

3 Results and discussion

3.1 Effect of HAL on chemical composition of the membranes

Based on the results of FTIR analysis (Figure 1.a.), the IR spectra of PSU and PSU/COM-HAL at different concentrations of COM-HAL were observed. For pure PSU, band peaks of the different functional groups present were determined having the following wavenumbers: 1013 cm^{-1} and 1103 cm^{-1} (aromatic C–H in-plane bend); 1148 cm^{-1} and 1171 cm^{-1} (O=S=O symmetrical stretch); 1239 cm^{-1} (C–O–C symmetrical stretch); 1292 cm^{-1} and 1322 cm^{-1} (O=S=O symmetrical stretch); 1488 cm^{-1} and 1586 cm^{-1} (C=C–C aromatic ring stretch); 2966 cm^{-1} (methyl C-H asymmetrical/symmetrical stretch).The band peaks in the IR spectra of PSU were all evident which confirmed the presence of PSU in the membrane. Moreover, the HAL IR spectra were also observed in accordance to the FTIR analysis by Bordeepong et al. (2011) [2]. The similarity between the obtained IR spectra for PSU/COM-HAL (Figure 1.a) and the expected IR spectrum for PSU/HAL (Figure 1.b) confirmed the interaction between PSU and HAL.

3.2 Effect of LiCl and HAL to the membrane surface morphology

Figure 2.b illustrates the SEM image of the electrospun PSU with the addition of LiCl, and as observed smooth fibers without beads and uniform fibers were produced as compared with pure PSU (Figure 2.a).

The obtained results confirmed that the addition of salt improved the conductivity of polymer solution which helps fiber formation significantly [1], but as observed, it also resulted to the decrease in the average fiber diameter from 697.37 ± 193.20 nm for pure PSU to 372.72 ± 130.50 nm for PSU/LiCl (Figure 3).

With regards to the addition of HAL to PSU/LiCl, curvy fibers with slight spindle-like bead can be observed with 0.5% COM-HAL (Figure 2.c) with an average fiber diameter of 521.78 ± 166.14 nm. Further increase in COM-HAL concentration (1% and 2%) showed no significant changes in the surface morphologies (Figures 2.d and 2.e), but it can be seen that the average fiber diameter somehow increases (Figure 3). For the addition of PH-HAL to PSU/LiCl, 0.5% and 1% concentrations resulted to the formation of spindle-like bead and bead-on-string morphologies (Figures 2.f and 2.g) with an average fiber diameter of 403.38 ± 99.000 nm. It can be seen in Figure 3, an increasing trend can also be observed for the average fiber diameter as the concentration of PH-HAL increases.

The SEM micrographs revealed that the average fiber diameter decreased significantly with the addition of LiCl to PSU membrane. Conversely, the incorporation of COM-HAL and PH-HAL on the membrane increased the average fiber diameter.

3.3 Effect of HAL to the wettability of the membrane

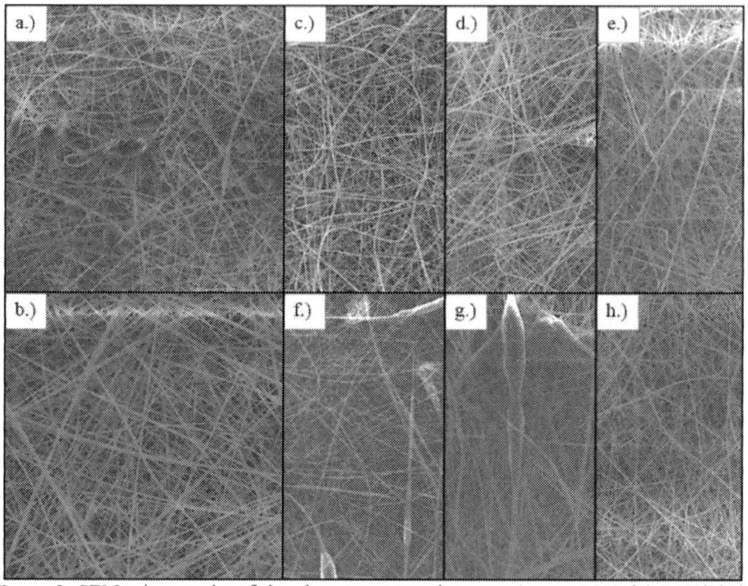

Figure 2. SEM micrographs of the electrospun membranes: a.) pure PSU; b.) PSU/LiCl; PSU/LiCl/COM-HAL at c.) 0.5%, d.) 1%, and e.) 2% COM-HAL concentrations; and PSU/LiCl/PH-HAL at f.) 0.5%, g.) 1%, and h.) 2% PH-HAL concentrations.

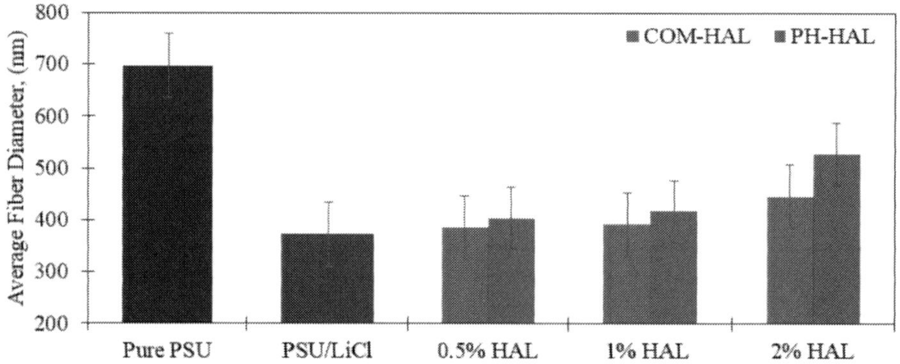

Figure 3. Effect of LiCl and HAL to the average fiber diameter of the membranes.

The contact angle of water on pure PSU was approximately 142 ± 2.94° which is in good agreement with the strong hydrophobic nature of PSU. Based on the contact angle measurements of water on the electrospun membranes, the addition of LiCl and HAL had no significant effect on the wettability of the membranes (Figure 4). The statistical analysis that was conducted, wherein p-value was greater than the significance level, also showed that hydrophobicity of the composite membranes was not affected by the incorporation of HAL. Although HAL is hydrophilic in nature, the increasing HAL content caused the surface of the nanofiber membranes to become rougher, which led to the decrease of the interfacial tension and surface energy [3]. Thus, the contact angle measurements lie within the range of each other.

3.4Effect of HAL to the mechanical properties of the membranes

As shown in Figure 5.a, increasing the concentration of PH-HAL decreases the tensile strength of the composite membrane. This is due to the weak interaction between PSU and HAL. The phenomenon was observed in the study of Peng et al. (2009) [4] in which they stated that the addition of nanoclay to the nanocomposites loss of tensile strength is due to the decreasing uniaxial orientation. Similar effect can be also observed with the addition of COM-HAL. Although there is a sudden increase in the tensile strength at 1% COM-HAL, the statistical analysis resulted to a p-value greater than the significance level suggesting that the addition of COM-HAL and PH-HAL had no significant effect to the tensile strength of the membranes. Moreover, it can be seen on the graph (Figure 5.b) that the addition of PH-HAL increased the maximum strain of the membrane as compared to the effect of adding COM-HAL. For the modulus (Figure 5.c), the incorporation of LiCl had no effect, but the values decreases with the addition of both COM-HAL and PH-HAL (0.5% and 2%). A sudden increase that can be observed at 1% concentration was due to the poor dispersion of the clay to the composite films resulting to high modulus. Considering the maximum tensile extension (Figure 5.d), the values increases with the concentration of COM-HAL, but no specific trend can be seen with the addition of PH-HAL. In general, decrease in the mechanical properties can be observed due to the effect of increasing HAL content which weakened the membrane; thus, decreasing the tensile strength [5]. However, upon statistical treatment of the data gathered, the p-values were found to be large and greater than the significance level which suggest that the addition of COM-HAL and PH-HAL at the concentrations used had no significant effect on the PSU membrane.

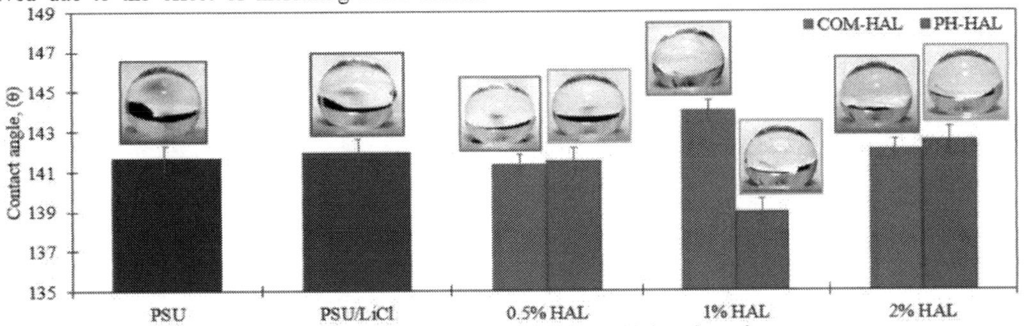

Figure 4. Contact angle measurements for the fabricated membranes

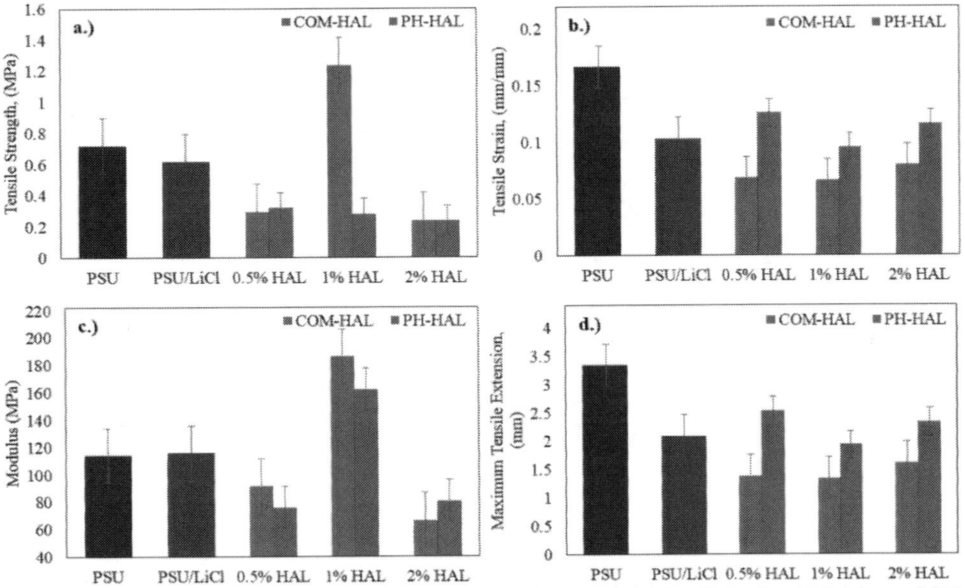

Figure 5. Mechanical properties of the electrospun membranes: a.) Tensile Strength, b.) Tensile Strain, c.) Modulus, and d.) Maximum Tensile Extension.

4 Conclusion

The researchers were able to synthesize nanofibrous membranes of PSU which contains COM-HAL and PH-HAL by electrospinning technique. It was observed form the SEM micrograph that the addition of LiCl salt produced bead free nanofibers with reduced average fiber diameters. The FTIR analysis verified the hydrogen bonding interaction of PSU and HAL with changing band peaks intensities from 3600 cm^{-1} to 3900 cm^{-1}. The addition of HAL on the membrane had no significant effect on the wettability of PSU. The composite membranes produced were evidently remained hydrophobic even with the incorporation of hydrophilic HAL. Also, the addition of COM-HAL and PH-HAL to the membrane decreased its mechanical strength. That is due to the exfoliation of the clay which optimized the number of available reinforcing elements; thus, increasing the matrix rigidity while often decreasing its fracture toughness [5]. However, the statistical analysis suggested that there was no significant difference in the values.

References

1. K. Chang and H. Lin, Electrospin of polysulfone in N,N'-dimethylacetamide solutions. *Journal of Polymer Research*, **16**, 611-622 (2009).
2. S. Bordeepong, D. Bhongsuwan, T. Pungrassami, and T. Bhongsuwan, Characterization of halloysite from thungyai district, Nakhon Si Thammarat Province, in Southern Thailand. Songklanakarin. *J. Sci. Technol.*, **33**, 599-607 (2011).
3. C. Luo, Z. Zou, B. Luo, W. Wen, H. Li, and M. Liu, Enhanced mechanical properties and cytocompatibility of electrospunpoly(L-lactide) composite fiber membranes assisted bypolydopamine-coated halloysite nanotubes, *Applied Surface Science*, **369**, 82-91 (2016).
4. Q. Peng, P. Cong, X. Liu, T. Liu, S. Huang and T. Li, The preparation of PVDF/clay nanocomposites and the investigation of their tribologicalproperties, *Wear,***266**(7-8), 713-720 (2009).
5. Y. Ma, F. Shi, Z. Wang, M. Wu, J. Ma and C. Gao, Preparation and characterization of PSf/clay nanocomposite membranes with PEG 400 as a pore forming additive. *Desalination*, **286**, 131- 137 (2012).

Density of test bodies and its effect on burning rate of fire-retardant treated samples

Linda Makovická Osvaldová

University of Žilina in Žilina, Faculty of Security Engineering, Žilina, Slovakia

Abstract. The thesis focuses on the effect of wood density on the results of fire tests. Wood (spruce wood in particular) is one of the most commonly used materials in many sectors - building constructions, furniture manufacture etc. With such a broad application, its fire technical properties must not be neglected. Fire retardants should be therefore used to prevent the fire from occurring and spreading.

1 Introduction

This physical property affects all technological processes as well as mechanical properties [4,1]. Therefore it has a direct impact on the production of components and parts for wooden constructions.

Wood density has an impact on other aspects of wood evaluation and its properties i.e. evaluation of fire-technical proprieties and retarding adjustments. Regardless of evaluation methodology (standard test methods or conical calorimeter) if the change in weight or thermal properties are used as the evaluation criteria, density may affect the final results of measurements. Test methods almost always specify the parameters for wood moisture, air conditioning, thickness or quality of working, however, the parameters for density (limit values at least) are missing. This article aims to identify the effect of density of spruce wood test specimens on the results of the experiments.

2 Material

Spruce wood (*Picea abies* (L) Karst Records.) is the second most widespread type of coniferous wood in Slovakia and the most important one from the economic point of view.

Different parts of tree (root, trunk and tree crown) have different functions and usability.

Chemistry in a live tree indicates chemical composition of necrotic cells (cell walls and lumens - inter-cellular spaces). From chemical viewpoint, wood is a complicated complex of heterogeneous bio-polymers (90 – 97 %) so-called major components accompanied by smaller quantities of accompanying components (3 – 10 %). Main components of wood are represented by saccharide share (65 - 75 %) - composed of cellulose (40 – 50 %) and hemicelluloses - plus aromatic share formed by lignin (15 – 35 %). Accompanying (accessory) components of wood are formed by organic substances - monomers and polymers - as well as by inorganic ones. In addition to organic substances, which are an essential part of wood, wood also contains mineral substances creating ash during combustion.

Cellulose is a high molecular polysaccharide consisting of Β-D- anhydro-derivate of beta-D glucopyranose (1→4).[1] For wood, average polymerization degree of natural cellulose evaluated viscosymetrically is 4000 up to 5500. One part of cellulose chains remains in amorphous arrangement while continuous transition between crystalline and amorphous share can be observed. Cellulose chains are maintained by hydrogen bonds between hydroxy groups of neighboring cellulosic chains.

Hemicelluloses are composed of heteropolysaccharides with lower polymerization degree (mostly 100 to 200). The main components are pentose and hexose: L- rahmnose, L- fructose, L- arabinose, D- xylose, D-mannose, D-glucose and D- galactoze. Hemicelluloses are amorphous and contain neutral or acidic cross-linked fibres. Coniferous plants contain glucomannans - the main hemicellulose component. Depending on the type of plant, their amount ranges from 20 to 30 %.[2] Lignin is a benzenoid component of some plants containing benzene core with propane chain, phenolic hydroxyl groups and forms a three-dimensional macromolecules.

Coniferous wood after thermal degradation is present by three changes – changes in chemical composition, in anatomical structure and in physical characteristic.

There are materials entering the reaction with oxygen (e.g. metals) or which, in the first stage of heating, change their chemical composition, or the composition of their core building elements, produce flammable gases and begin to burn. Wood belongs to this category too. Main components of wood degrade at different temperatures affecting the whole process of combustion. Hemicelluloses will decompose under temperatures ranging between 170 – 240 °C. They are the least resistant to thermal decomposition. By regulating wood heating, we may achieve reduction of hemicelluloses without reducing its strength, resulting in an increase in dimensional stability of wood. [3]

Cellulose is more resistant to thermal stress than hemicelluloses. Decomposition rate of cellulose is moderate up to 250 °C. Intensive decomposition of cellulose is observed at the temperatures of 250 -300 °C. It leads to disruption of links in the main chain according to radical mechanism, the tail link of cellulose turns into levoglucosan.[3]

Lignin is the most resistant to thermal decomposition. Process of the thermal decomposition of lignin is divided into two stages. In the first stage, labile ether alkyl links decompose at the temperatures of 300 - 320 °C. The second stage is an active decomposition of lignin at the temperatures of 350 - 390 °C. The structure of macromolecules is disrupted and volatile products are being released. Formation of volatile products is lower than with cellulose. Decomposition rate is slowing down by increasing the temperature and is followed by accumulation of condensing aromatic structures in the solid phase.[4]

Wood exposed to temperatures higher than 288 °C (temperature generally assumed as charring temperature) can be divided into five degradation zones.[4] The variety of the decomposition of main wood components and characteristic reactions in different temperature intervals lead to differentiation of burning process into several stages.

Anatomical structure of wood - individual cell elements - affects the process of combustion, its first phase – ignition - in particular. It results mainly from chemical composition and geometric shape of cellular elements, their size and their number.

In addition to chemical composition, physical properties of wood and wood-based materials significantly influence the course of combustion. Each physical property has an impact on burning, though not to the same degree. [5]

Surface of the material (its quality) is another physical characteristic, which significantly affects combustion. Wood, capillary-porous material, is rough on the surface. The roughness depends on working as well as on anatomical structure of wood. Besides roughness, surface quality also depends on anatomical defects, defects resulting from working, mechanical damage, dirt and etc., changing the surface quality. Surface quality mainly affects thermal conductivity α. Smooth and high-quality surface reflects energy of radiation source and flame source thus making it more difficult for wood to be ignited compared to rough surface under the same test conditions.

From the point of view of quality, working is as important as the color of wood. Natural color of wood (including tropical woody plants) comprises the entire color range. Thermal degradation causes color changes. Light color of wood gradually darkens and finally turns black - charred layer. Intensity of darkening depends on heat rate and temperature - several authors treated the subject. Color change is brought about by polysaccharide degradation (hemicelluloses and cellulose). [5]

Test specimens made from spruce wood are sorted out according to their density. Out of 450 specimens, 60 pieces have been divided into three density groups (see Table 1).

2.1 Selection criteria

2.1.1 Density

$$\rho = m/V \qquad (kg/m^3) \tag{1}$$

where:

ρ – density (kg/m^3)

m – weight (kg)

V – volume (m^3)

According to the following formula, density of the selected samples has been calculated. Samples have been then arranged from the ones of lowest to highest density. All samples have been then divided into three categories. Samples of the lowest densities with the average value of 381.5 kg/m^3 have been included into low density group. Samples of medium values have been included into medium density category. The average values of all samples were 437.87 kg/m^3. Samples with the highest values have been included into the category of high density - their average density value was 495.62 kg/m^3. Each category was divided into two groups - fire-retardant treated samples and untreated samples. Their average values are shown in table 1.

Table 1. Average densities of samples.

Density (kg/m^3)	Average density of untreated samples	Average density of fire-retardant treated samples
Low density	378.0	385.0
Medium density	438.4	437.4
High density	492.4	498.9

2.1.2 Fire retardant

Fire retardant called Ohňostop was randomly selected for the experiment. The retardant will not be described or characterized in details. The goal of the experiment was not to evaluate the retardant. Our goal is to find out whether the density of test bodies has an effect on the overall result even with fire retardant treated samples.

2.1.3 Thermal load

The method of heat load application is based on standard test method where weight loss is the main assessment criterion. Test specimens of 200 x 100 x 10 mm (10 mm - thickness) have been exposed to the flame of a gas burner for 10 minutes. The flame has been directed at the middle of the test specimen which formed 45° angle with the horizontal plane. Perpendicular distance of the center of the sample and the mouth of the burner was 90 mm. Flame is 100 mm long.

2.2 Evaluation criteria

Relative burning rate has been determined according to the following relations (2) (3):

$$v_r = \left|\frac{\partial \delta_m}{\partial \tau}\right| \quad (\%/s) \qquad (2)$$

or numerically

$$v_r = \frac{|\delta_m(\tau)-\delta_m(\tau+\Delta\tau)|}{\Delta\tau} \quad (\%/s) \qquad (3)$$

where:
v_r – relative burning rate (%/s),
$\delta_m(\tau)$ – relative weight loss in certain time (τ) (%),
$\delta_m(\tau+\Delta\tau)$ – relative weight loss in certain time ($\tau + \Delta\tau$) (%),
$\Delta\tau$ – time interval at which weights are being subtracted (s).

3 Evaluation

Labels were as follows: NNX for untreated samples and RNX for fire-retardant treated samples of low density; NSX for untreated samples and RSX for fire- retardant treated samples of medium density; NVX for untreated samples of high density samples and RVX for the samples treated with a flame retardant.

Maximum relative burning rate of 0.0958 s/% for NNX samples was achieved in 60th second. Maximum relative burning rate of 0.0476 s/% for RNX samples was recorded in 80th second. These values of relative burning rate of treated and untreated samples of low density show significant differences between the two groups proving the importance of fire-retardant application. These differences can be seen from figure 1.

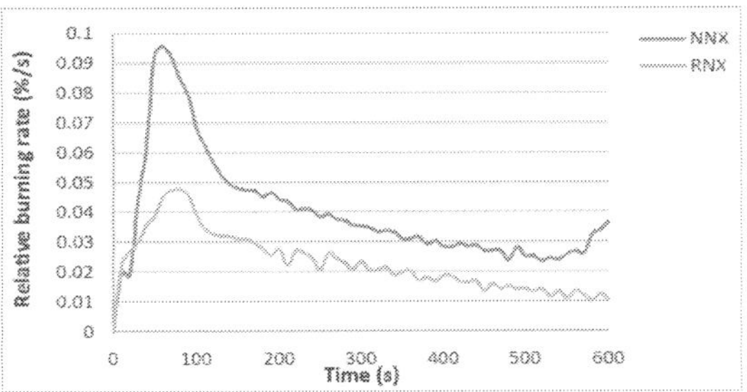

Figure 1. Average relative burning rate of low density samples.

Maximum burning rate for RSX samples was 0.0496 s/% in 60th second. We observe greater differences between the values of relative burning rate of medium density samples treated with fire retardant and untreated ones compared with low-density samples. These differences are shown in figure 2.

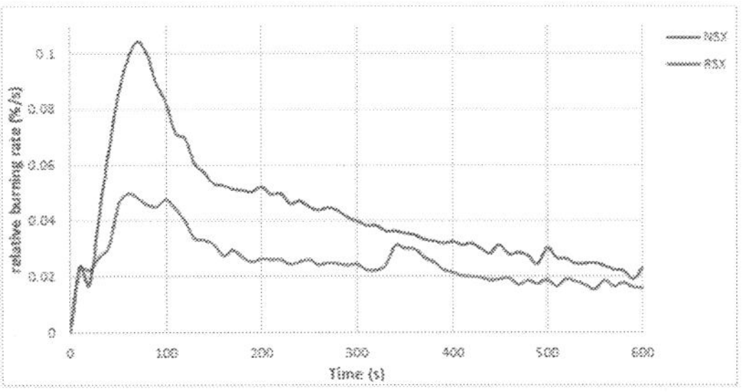

Figure 2. Average relative burning rate of medium density samples.

Maximum relative burning rate for NVX samples was 0.0862 s/% in 80th second. Maximum relative burning rate for RVX samples was 0.0512 s/% in 70th second. The smallest difference can be observed between these values - weight loss and relative burning rate of treated and untreated samples of high density - compared to samples of low and medium density. These differences are shown in figure 3.

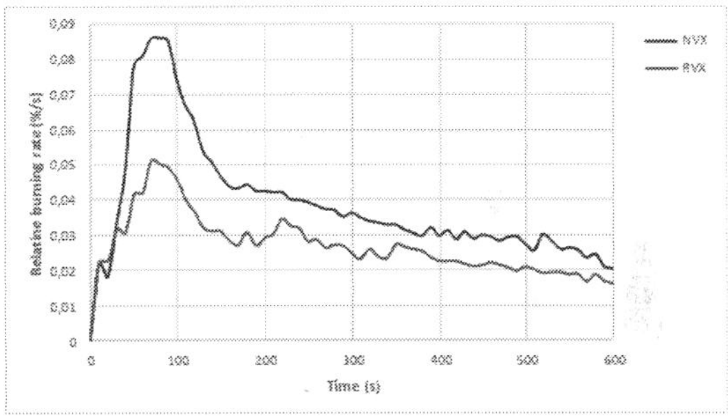

Figure 3. Average relative burning rate of samples of high density.

3.1. Overall evaluation of laboratory tests

Overall evaluation of laboratory tests has been done using all the average values, i.e. average values of untreated samples of low density (NNX), fire retardant treated samples of low density (RNX), untreated medium density samples (NSX), medium density samples treated with a fire retardant (RSX), untreated high density samples (NVX) and high density samples treated with a fire retardant (RVX).

Based on the values shown in the table 2 and fig. 4, following facts can be concluded:
- Fire retardant played an important role since the untreated samples of low, medium and high density

suffered greater weight loss than the samples of low, medium and high density treated with a fire retardant.
- The greatest difference between the treated and untreated samples occurred in the case of medium density samples since these samples have absorbed the largest quantity of fire retardant.
- On the other hand, the smallest difference between the treated and untreated samples occurred in case of high density samples since they absorbed the lowest quantity of fire retardant.
- The greatest weight loss was recorded for untreated samples of low density.
- The lowest weight loss was observed in the case of samples of medium density treated with a fire retardant.
- Each group of untreated samples has achieved higher maximum relative burning rate than the untreated ones.
- The highest relative burning rate has been observed with the untreated samples of medium density since the value was achieved 10 seconds later than with the untreated samples of low density that achieved the second highest relative burning rate.
- The lowest relative burning rate has been recorded for fire-retardant treated samples of low density.

Table 2. Evaluation of laboratory tests.

Density - label	**Relative burning rate (%/s)**
NNX	0.0958
RNX	0.0476
NSX	0.01042
RSX	0.0946
NVX	0.0862
RVX	0.0512

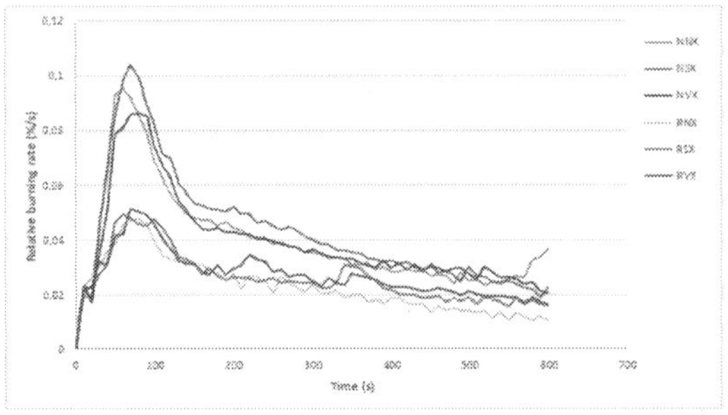

Figure 4. Relative burning rate of average values (all samples).

4 Conclusion

In conclusion - density of wooden test bodies has a direct effect on the results of measurements of retarding effect of fire retardants. The experiment monitored the evaluation criteria related to weight or its change during the evaluation of fire retardants. Experiment confirmed that the more retardant is applied, the lesser is the effect of density on test results, and however, this aspect cannot be neglected. Moreover, if quality selection of test specimens - according to their density before the treatment/experiment - is not carried out, a great variability of results in the given group can occur. The auxiliary statistical criteria may affect the overall evaluation of the given fire retardant. We should point out that density will have an effect on the evaluation criteria which aim to track the changes in temperature or the effect of heat flow [6,7,8].

Acknowledge

This work was supported by the Slovak Grand agency VEGA. (Project 1/022/16|6| Fire safe insulation systems based on natural materials.).

References

1. Osvald A, Makovická Osvaldová L, Kačíková D. Coniferous wood - reaction on fire. In: Wood and Fire Safety; 13. - 16. May 2012; The High Tatras Štrbské Pleso, Slovakia. 2012. p. 203-210.
2. Bubeníková T, Kačík F, Kačíková D. Characteristic of lignins at low temperature degradation of spruce wood. In: Proccedings Wood and Fire Safety; 2004; Štrbské Pleso, Slovakia. Zvolen, Slovakia: Technická Univerzita; 2004. p. 25-30.
3. Hawley LF. Combustion of wood. Wood chemistry. 1952;**II**:548.
4. White RH, Nordheim EV. Charring Rate of Wood for ASTM E 119 Exposure. Fire technology. 1992;**28**(1):5-30.
5. Martinka, J., Chrebet, T., 2014 a: Activation energy of Teak and Oak wood spontaneous ignition. In: Advanced Materials Research, Switzerland, vol. 1001, 2014, pp. 262-266.F. De Lillo, F. Cecconi, G. Lacorata, A. Vulpiani, EPL, **84** (2008)
6. XU Qiang, QUE Xinggui, CAO Liying, JIANG Yong, JIN Cong, 2010: Study of heat flux distribution of wood crib fires: bench scale tests, Thermal Science, 2010, Vol. 14, No.1,pp 283-290
7. Tsuomis, G., 1991: Science and technology of wood; Structure,properties, utilization. NewYork :Chapman and Hall, p. 494. ISBN 0-412-07851-1
8. Kaľavská Ľubomíra : [Diploma thesis].University of Žilina. Faculty of Security Engineering;Department of fire engineering, study programme: Rescue services. 2016, 62 pages

Simulation and Experimental Study on Flow and Heat Transfer of Ship Central Cooling System

Fang Chao , Cai Biao-hua , Ma Shi-hua and Li Bin

Wuhan Second Ship Design and Research Institute, Wuhan 430064, China

Abstract. The centralized cooling system of a certain type of oil ship is a large-scale complex fluid network system. Fluid system simulation software (FLOWMASTER) was used to establish simulation model for steady operating conditions. The design of central cooling system was validated reasonably by simulation in steady-state condition. In this paper, the flow and heat transfer experiment was carried out. The experimental results were in good agreement with the simulation results, which suggested that using FLOWMASTER to validate the heat transfer performance of fluid system is reliable.

1 Introduction

The marine centralized cooling system provides and distributes seawater and freshwater to multiple users, ensuring the onboard air compressor and inverter to operate in a safe temperature range. The design of ship cooling system has certain complexity. Reasonable choice and design of cooling system have important influence on ship`s economy and seaworthiness [1].In the traditional ship cooling system design and debugging, design is mainly based on empirical formula, and debug is usually relied on engineering experience[2]. If the test conditions are available, the rationality of the system design would be verified by experiment.

In this paper, according to the pipeline design of the centralized cooling system of a certain type of tanker, a simulation model was established by using the fluid system simulation software FLOWMASTER. On the basis of modeling, the steady flow and heat transfer calculation of the pipeline system were carried out. The temperatures and flow parameters at key positions of the system were compared with the designed values to verify the rationality of the design .At the same time, the flow heat transfer experiment of the central cooling system was carried out, and the experimental results are in good agreement with the simulation results.

2 Overview of marine centralized cooling system

As shown in Figure 1, the centralized cooling system of a certain type of tanker consists of two parts, a closed loop freshwater cooling circuit and an open loop seawater cooling circuit. The closed loop freshwater circuit cools the air compressor, inverter and other equipment. The open seawater cooling circuit uses low temperature seawater overboard to cool high temperature freshwater centrally through intercooler. The high temperature seawater was discharged overboard after the heat exchange.

The freshwater cooling circuit provides cooling water for several users, including 3 air compressors located in different compartments, and 4 inverters located in the same cabin. The total flow demand of freshwater is supplied by a single freshwater pump, and the chilled flow of each user is regulated by the outlet flow control valve of freshwater pump, the bypass flow control valve and the user branch flow control valve. In the seawater cooling circuit, a single seawater pump is used to supply seawater to 2 series central coolers. According to the temperature of the seawater and the heat load of the freshwater, the supply of low temperature seawater is regulated by the bypass flow regulating valve and outlet regulating valve of the seawater pump.

The design idea of the separation of freshwater circuit and seawater circuit in marine centralized cooling system can fundamentally reduce the corrosion of most pipelinescaused by seawater in ships. And it is the development trend of ship cooling system in the future. However, the centralized cooling system has become more complex, and the design, commissioning and checking work has changed considerably. In this paper, the rationality of system design is verified by simulation and analysis of flow heat transfer in centralized cooling system, which is of great significance for shortening design cycle and reducing design cost.

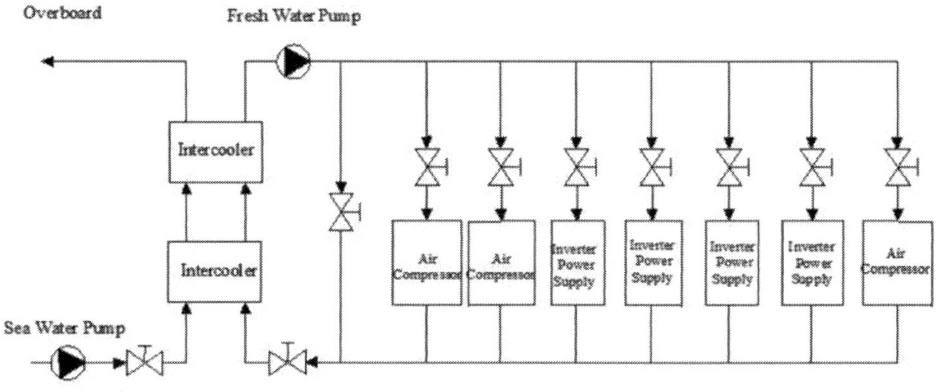

Figure 1. Schematic diagram of a certain type of ship centralized cooling system

3 Modeling of marine centralized cooling system

3.1 Fundamentals of fluid networks

FLOWMASTER is one-dimensional fluid system simulation software for engineering applications. For complex pipeline systems in various engineering, a system model can be established quickly and effectively to be analyzed completely in FLOWMASTER [3-7]. This software can analyze the fluid network system in steady state and transient state, and can also analyze the flow and heat transfer of the fluid network system and the coupling analysis between the two systems [8].

The numerical calculation of the software is based on fluid network analysis method. Fluid network analysis simplifies the fluid system by connecting a series of nodes and pipe sections. The following assumptions are made for the nodes:(1)The node has a certain volume and the fluid in the node is in a homogeneous state;(2)The storage of fluid mass can only occur in the interior of the node, and all flow resistance of the system is concentrated on each branch.

The pipe network is composed of C pipe sections and N+ 1 node, and the mathematical model of the pipe network is as follows:

$$B \cdot G = dM / dt \tag{1}$$

$$B^{'} \cdot P = S \cdot |G| \cdot G + \rho \cdot gZ - DH - L \cdot \frac{dL}{dt} \tag{2}$$

$$\overline{B} \cdot q - \underline{B} \cdot (q + r) = dH / dt \tag{3}$$

Among them, B is the $N \times C$ order correlation matrix of the pipe network; G represents branch traffic, C order vector; M represents node quality, N order vector; $B^{'}$ is the $C \times N$ order correlation matrix of the pipe network, the transpose matrix of B; S is the resistance coefficient of each other, $C \times C$ order diagonal matrix; Z is the height difference of the pipe section, C order vector; DH represents the boost of the power source on each branch, C order vector; L represents flow inductance, numerically equal to the amount of pressure caused by 1 units of flow change; P represents the pressure at the node; ρ represents the density at the node;g represents the acceleration of gravity; $\overline{B} \cdot q$ represents the sum of heat flow of each branch to the node; $\underline{B} \cdot (q + r)$ represents the sum of heat flow at the outflow node; H represents the enthalpy of each node.

Equation (1)、(2)and(3) are the concrete expressions of fluid dynamics, continuity equations ,momentum equations and energy equations in one-dimensional pipe networks. FLOWMASTER is a kind of software for simulation analysis of fluid system. The essence is to substitute the relevant parameters of the system components into the equation, obtaining the flow, pressure, temperature and other parameters.

3.2 Physical simulation model

According to the actual design of the centralized cooling system of the oil ship, a one-dimensional simulation model of the cooling system pipe network is established by using FLOWMASTER software. The pipe network includes pipes, pumps, tee coupling, valves, etc. The overall flow loss of centralized cooling system is large, so the influence of low flow resistance components such as elbow is not considered in the modeling process. At the same time, to reduce the number of pipes and tee couplings, discrete loss components are used to replace a large number of flow resistance components in the system, so as to improve the computational efficiency. The simplified simulation model is shown in Figure 2.

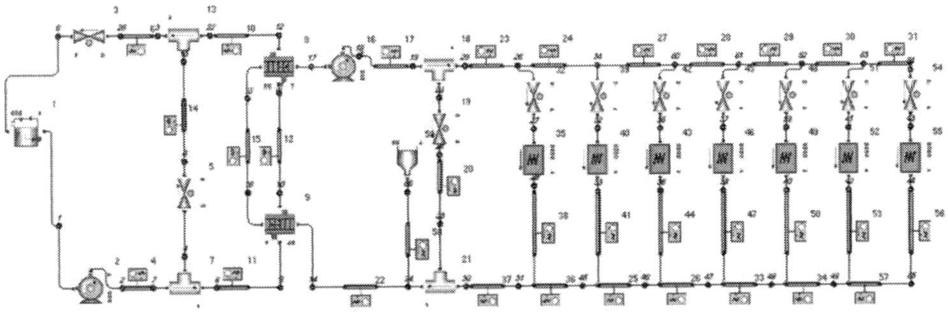

Figure2. Centralized cooling system piping model

4 Steady state simulation of flow heat transfer in concentrated cooling system

4.1 Simulation conditions and parameter settings

On the basis of flow simulation modeling of centralized cooling system,the steady state simulation of flow heat transfer in centralized cooling system was carried out by taking common navigational conditions as an example. Under this condition, 3 sets of air compressor in different compartments all are put into operation and 4 sets of inverter in the same cabin work at the same time. The cooling system adjusts the user traffic through the distributor valve, branch valve and bypass valve to meet the cooling demand. The flow demand and heat load of the fresh water cooling system under common navigation conditions are as follows.

Table 1 User flow demand and heat load under common navigation condition

Order	Cooling user name	number	Single flow demand (m^3/h)	Single thermal load (kW)	Cooling medium	Use attributes
1	air compressor	3	7	51	Fresh water	Common
2	inverter power supply	4	2.5	25	Fresh water	Common

You are free to use colour illustrations for the online version of the proceedings but any print version will be printed in black and white unless special arrangements have been made with the conference organiser. Please check with the conference organiser whether or not this is the case. If the print version will be black and white only, you should check your figure captions carefully and remove any reference to colour in the illustration and text. In addition, some colour figures will degrade or suffer loss of information when converted to black and white, and this should be taken into account when preparing them.

4.2 System design check

According to the design parameters of the centralized cooling system, the centralized cooling system model was simulated in steady state by regulating the detour valves, branch valves and bypass valves. By comparing the simulation results with the design values, the rationality of the centralized cooling system design was checked. As shown in the following table:

Table Import and export parameters of fresh water and sea water of central cooler under common navigation conditions

Parameters	Fresh water inlet		Fresh water outlet		Sea water inlet		Sea water outlet	
	Design value	Simulati-on value	Design value	Simulati-on value	Design value	Simulati-on value	Design value	Simulati-on value
Temperature/℃	38	38.13	—	34.23	28	28.21	—	33.22
Pressure/bar	—	0.28	—	0	—	1.52	—	1.51
Flow speed/$m \cdot s^{-1}$	—	1.74	—	1.74	—	0.40	—	0.40

The design specifications of the centralized cooling requires that the temperature difference between central cooler's freshwater outlet and heat exchanger's seawater inlet shall not be greater than 10 degrees centigrade. The simulation results show that the fresh water outlet temperature of the central cooler is 34.23 ℃ and the seawater inlet temperature of central cooler is 28.21℃。Therefore, the temperature difference between the freshwater outlet of central cooler and seawater inlet of heat exchanger is 6.02 ℃ , which satisfies the requirement of the design instructions that the temperature difference between the two is not greater than 10 ℃。Therefore, the design of the centralized cooling system of the ship is reasonable.

4.3 Comparative analysis of experiment and simulation

In order to prove the rationality of the system design checked by using FLOWMASTER simulation, the heat transfer experiment of the centralized cooling system was carried out, and the critical position flow, pressure, temperature and other parameters were obtained.

The temperature of the sea system is 28.07℃ and the laboratory air temperature is 20 ℃。Heat transfer performance experiment was carried out on common sailing conditions. First, start the freshwater pump, and adjust each cooling branch flow of fresh water circuit to make the closed circuit of fresh water operate steadily. Then, start the seawater pump and adjust the flow of the system to make the seawater loop run stably. Finally, start the all heat source simulation devices,and adjust the 3 air compressors' heat power to 51kW and 4 inverters' heat power to 25kW respectively. The inlet temperature, the fresh water outlet temperature and the user flow of the centralized cooling system were measured. The results are shown in Table 3.

Table3 Experimental results of heat transfer in common navigation condition

Order number	T1	T2	H1	H2	H3	H4	H5	H6	H7
1	27.51	29.6	7.35	7.25	7.24	2.61	2.56	2.56	2.59
2	27.51	33.89	7.33	7.25	7.19	2.58	2.53	2.54	2.61
3	27.51	35.1	7.26	7.28	7.20	2.56	2.51	2.53	2.60
4	27.84	35.76	7.19	7.26	7.17	2.56	2.50	2.51	2.61
5	28.07	35.99	7.19	7.23	7.17	2.54	2.50	2.51	2.61
6	28.28	36.31	7.22	7.26	7.19	2.54	2.50	2.51	2.61
7	28.61	36.64	7.19	7.26	7.06	2.54	2.50	2.51	2.61
Mean value	27.90	34.76	7.25	7.26	7.17	2.56	2.51	2.52	2.61
Simulation value	28.21	34.23	6.92	7.11	6.98	2.48	2.43	2.44	2.49

Note:T1-Temperature of central cooler seawater inlet,T2-Temperature of central cooler fresh water outlet,H1～H3-Air compressor cooling flow,H4～H7-Inverter cooling flow.

The experiment results show that the difference between the temperature of fresh water outlet and the temperature of seawater inlet is 6.86 ℃,and it meets the design requirements as well. The simulation results of key position of temperature and flow rate are consistent with the experiment results, and the error is in the allowable range. So the simulation can reflect the heat transfer performance of the ship central cooling system. Therefore, the simulation analysis based on FLOWMASTER software to verify the rationality of the design is reliable, and the software can be applied to the design, commissioning and verification of the pipe network system.

5 Summary

In this paper, FLOWMASTER software was used to carry out steady state simulation calculation of a certain type of ship centralized cooling system, and the temperature, flux and other parameters of the critical position were obtained. According to the design specification, the rationality of system design was checked. The results show that the centralized cooling system meets the design requirements. In order to further prove the reliability of FLOWMASTER system simulation, the heat transfer experiment of the centralized cooling system of the ship was carried out. Comparing the experimental results and simulation results, the results show that the system simulation using FLOWMASTER can reflect the heat transfer performance of the system effectively and truly.

References

1. ZHENG Xuan-liang, Liu Xi-wei, Zeng Qing-qian. Heat balance calculation and program simulation design of ship central cooling system[J].Naval Architecture and Ocean Engineering, 2012(2), 42-57.
2. XU Peng, Fan Hao. Simulation of Bilge Water System in a Ship Based on FLOWMASTER [J].Ship & Ocean Engineering,2016,45(2),86-89.
3. ZHANG Guang-peng, Xu Nuo, Zhang Wu-ping. Application of FLOWMASTER in HVAC[J].Refrigeration and Air Conditioning, 2006(3),34-36.
4. HOU Yu-zhu, ZHANG Yuan-suo,WEN Chao-zhu, DONG Wei. Analysis and research of the pipeline flow simulation in central cooling system of naval vessel[J].Ship Engineering,2009(31),75-76.
5. LIU Zhen-dong. Professional Fluid System Simulation Software FLOWMASTER [J].CAD/CAM and Manufacturing Information Industry.
6. MEI Xing-xin, TANG Ling-di, TANG Yue. Application of FLOWMASTER in Hydraulic Balance of Air Conditioning Chilled Water

System[J].Heating Ventilating & Air Conditioning,2014(4):92-95.
7. SHI Hong, JIANG Yan-long, LIU Zhi-li, etc. Simulation of Bleed Air Behavior During Air Craft in Flight Based on FLOWMASTER [J] .Transactions of Nan-jing University of Aeronautics & Astronautics, Jun, 2013(2) : 132-138.
8. D S Miller. Internal Flow System [M].BHRA Fluid Engineering,1978.

MATEC Web of Conferences **213**, 04002 (2018) https://doi.org/10.1051/matecconf/201821304002
ACMME 2018

A statistical Approach for Finding Influential Factors in Respect of Energy Consuming of A Car Passenger

Hussein M. Ali[1] , Asif Iqbal[2]

[1] *Ph.D. in Mechanical Engineering; Tel: +9647708372370; Email: alabadi.hussein@ntu.edu.iq & hrlha@yahoo.com, Department of Refrigeration and Air conditioning Technology Engineering, Technical Engineering College of Mosul, North Technical University, Mosul, Iraq*
[2] *Ph.D. in Mechanical Engineering; Tel: +673- 8929497; Email: asifiqbal@gmail.com Faculty of Integrated Technologies, University Brunei Darussalam, Jalan Tungku Link, Bandar Seri Begawan BE1410,Brunei Darussalam*

Abstract: Nowadays the usage of gasoline as an energy resource is one of the most important subjects in the engineering field. A car is one type of energy consumer. Energy is used to build the cars and to running it. The fuel prices are fluctuate, so it seems sensible to explore every avenue towards saving energy in cars making. and study the factors that affect its consumption. The aim of the present work is to explain theoretically the calculation of fuel saving and cost in a car passengers in a greater detail than it has been done before and to describe statistically the affecting factors upon it. A statistical analysis has been used to study the influence of the weight and acceleration of the car upon the fuel consumption. It was shown that the fuel consumption increases linearly with the increase of a car weight and accordingly, the cost per unit travel of the car will increase.

Nomenclature:

s.f.c Specific fuel consumption
Fw Attractive force.
Fd Drag force.
FR Friction rolling
Fc Climbing force.
g Gravity acceleration
$\overline{\rho}$ Density of the air
C w Drag coefficient
V Velocity of the car (m/sec)
V_0 Velocity of air (m/sec)
Pw Motive power to resist attractive force
S Travel distance(m)
t Time (sec)
F Total force which run the car
m Mass of the car (kg)
a Acceleration (m/sec2)
ρ Density of the fuel
Q Volume of the fuel
S/Q Unit travel of the car per unit volume of fuel (Km/Lit or Mile/gallon)
η Transmission system efficiency

1 Introduction

Several crises in the world had been resulted in the disappearance of cheap fuel oil. During the last 40 years, there have been many oil price shocks. After these shocks the world had reacted with recession and economies so that, the price was sliding up to unknown levels. Energy is used to build a car and to run it. The energy spent in materials, factories and on the roads. Examination shows that, the energy used to move the cars dominates the energy to produce cars (15% of the total energy consumed by nation table 1[1]. With a limited supply of traditional non-renewable energy and a slowly growing renewable energy industry, it is important to increase energy efficiency to meet energy demand. Therefore, it is seem sensible to explore every avenue towards saving the energy used in making & running cars. Many researchers have carried out the study of the affecting factors on the fuel consumption of automobile. Al-Momani et al (2007)[2], studied experimentally the effect of different factors (tires' conditions, altitudes, passengers' weight and the use of air-conditioning unit (load), vehicle speed, fuel type, maintenance condition) on the vehicles' fuel consumption. They found that the significant improvements in automotive fuel economy can be obtained by reductions in load, aerodynamic drag and rolling resistance (tire improvements), as well as by improvements in maintenance conditions and using efficient fuel type. Charles et al (2013)[3], develop a theoretical model for the rate of fuel consumption of a vehicle (especially of a highway vehicle) as a function of its airspeed, taking into account the fuel consumption due to the moving parts of the vehicle as well. They found that the theoretical model is correctly describe the experimental data on fuel economy of a vehicle as a function of airspeed. The theoretical model as well as the experimental data clearly show that the fuel economy of a vehicle does indeed depend on its air speed. Uldis (2011)[4], measured the load as one factor that affect the fuel consumption of a car. He found that the fuel consumption is increasing by increasing the load, every 100 kg load increases the fuel consumption by 2.58-2.95 % varying from the driving cycle in on-road, and 2.57- 2.89 % on chassis dynamometer experiments. Jan Rouwendal (1996)[5], analyzed the fuel efficiency of private cars in relation to both the technical characteristics of the cars and the socio-economic characteristics of a driver. This effect has been investigated for a sample of Dutch drivers. He found that the age, type of employment of the driver and the price of the gasoline have significant effects upon the fuel efficiency.

In the present work, a statistical analysis has been used to study the influence of the weight and acceleration of the car upon the fuel consumption. For this purpose, the following guide lines have been followed:

2 Pre-Experimental Planning

2.1Recognition of system

For the present study, a car (Golf model 1.3 trim C.CL.GL) of the following technical data has been used [6], is:
Capacity (1227 cm3)
Weight of the car = 751.9 Kg
S.f.c – 0.34 Kg/KW.hr = 9.68*10-8 Kg/w.sec
ρ = 780 kg/m3
Face area = 2.2795 m2
Acceleration =2.1 m/sec2

$f = 0.015$

$C_w = 0.34$

$\eta = 0.9$

$V = 25$ m/sec

$V_0 = 2.77 m/\sec$

2.2 Theoretical Analysis:

To understand the factors involved in running the car, the following mathematical analysis have been carried out as shown in Fig. 1.[3], .

$F_w = F_c + F_R + F_d$

where:
Fw = attractive force.
Fd = Drag force.
FR = Friction rolling = m * f * g
Fc = climbing force.

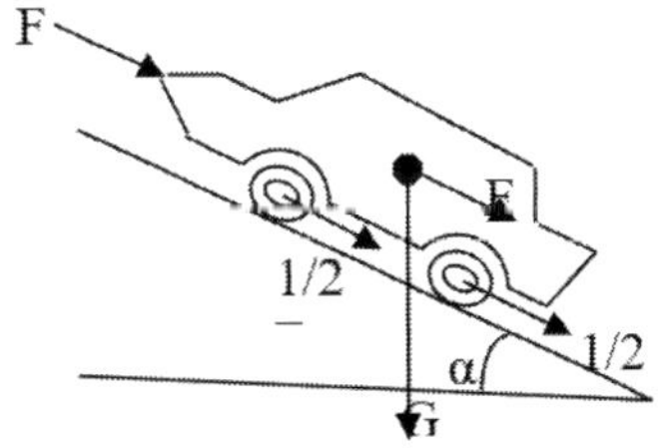

Fig. 1 forces effecting in running the car

After some derivatives for the elements of the above relation both sides of the above equation from zero to t. the following relation was obtained:

$$\frac{\eta * Q * \rho}{s.f.c} = s(F_w + m * a)$$

$$\frac{Q}{s} = \frac{(F_w + m * a) * s.f.c}{\rho * \eta}$$

$$\therefore \frac{s}{Q} = \frac{\rho * \eta}{(F_w + m * a) * s.f.c}$$

where:

$\frac{s}{Q}$ = unit travel of a car/unit volume of fuel

As shown in the equation above:
ρ, η and s.f.c Can be regarded as constant

Therefore $\frac{s}{Q}$ is a function of a mass of car (m),

acceleration (a) and attractive force (F_w).
But:

$F_w = F_c + F_R + F_d$

$F_R = m * f * g$

$F_d = 0.5\bar{\rho} C_w A(\upsilon + \upsilon_0)^2$

i.e. $\frac{s}{Q}$ vary with $m, a, f, C_w, A, \upsilon, \upsilon_0$

2.3 Choice of factors and selection of the response variable

In selecting the response variable, we have to be certain that variable completely provides clear, correct and useful information about the process or system that under study. In this experiment, the fuel consumption of a car has been chosen as the response that has to be reviewed. Now, the question is what are the factors that effect upon the fuel consumption of any car?

Actually, there are many factors (controllable, uncontrollable, or noise factors) effect on the fuel consumption of a car, these are: (Weight of the car, Road properties., Maximum speed, Acceleration of the car, Drag force affected by the temperature and air pressure, Aerodynamic shape of a car and Velocity of the wind, etc.). In the present work, it is interested to study the effective of the weight and acceleration of the car that have more effective two factors upon the value of the usage of gasoline by the car. Each of them has special levels. We choose those two factors so we can say that there are fixed effect factors. In the first view, it seems that the weight of the car is constant, but it's not true. Because the weight of car is the summation of pure weight of empty car and what is inside in it, including persons and other things in it. Then by changing the car weight from a minimum value (pure empty car weight plus a minimum assumed as driver's weight) to the maximum value of weight that can be held by the car we can find the limitations of the car weight as the factor that must be studied.

The data of this experiment have been chosen randomly and can be shown on the table 1.

Table 1 Value of S/Q for each experiment. The values in the dark cells are average of each grand cell. The Values unit is (km/lit)

		acceleration (m/s^2)							
		2.1		3.1		4.1		5.1	
weight (kg)	827	4.037	4.19	2.940	3.381	2.312	2.421	1.905	2.06
		4.356		3.821		2.529		2.225	
	902	3.752	4.22	2.723	3.223	2.137	2.153	1.758	2.22
		4.691		3.722		2.168		2.683	
	977	3.504	3.85	2.535	2.694	1.986	2.480	1.632	1.87
		4.207		2.852		2.973		2.125	
	1052	3.287	3.37	2.371	2.408	1.855	1.922	1.523	1.600
		3.454		2.444		1.988		1.676	
	1127	3.096	3.31	2.228	2.289	1.740	2.099	1.427	1.71
		3.525		2.350		2.458		1.998	

As shown in table 1, for each situation of the two factors, because of lack of resources and expensiveness of each experiment, we just did two experiments.

In table 1 there are (40) data observed randomly and all of them it will be use. By choosing the response, data, and factors we can be sure about first and important part of the experiments. This means that, the pre-experimental planning is complete.

3 Experimental Design

3.1 Choice of experiment

If the pre-experimental planning activities are done correctly, the choose of experiment becomes relatively easy. In this section we have to select a number of replication and determine which type of design should be used. There are several designs that can be choose to do this experiment but, because of the lack of the resources and expensiveness of the experiments, we have to choose the number of replication as two replicates under each weight and acceleration.

3.2 Factorial Design

In this paper, we suspect that there are two factors effect on the response of experiments (fuel consumption of the car) more than the other factors. There are many design experiments involve the study of the effects of two or more factors. In general, factorial designs are most efficient for this type of experiment. By a factorial design, in each replication of experiment, all possible combination of the levels of the factors was investigated[7].

Because of the above reason, we had used factorial design of choice of experimental design in this experiment, and as result the two-factor factorial design applied for experiments. Table 2 shows the basic design of two-factor factorial design.

As shown in table 2 there are two factors, acceleration as factor and has five levels and weight as factor B that has four levels (based on table 1).

Table 2 basic design of two-factor factorial design

		Factor A							
		1		2		3		4	
Factor B	1	y_{111}	$\bar{y}_{11.}$	y_{121}	$\bar{y}_{12.}$	y_{131}	$\bar{y}_{13.}$	y_{141}	$\bar{y}_{14.}$
		y_{112}		y_{122}		y_{132}		y_{142}	
	2	y_{211}	$\bar{y}_{21.}$	y_{221}	$\bar{y}_{22.}$	y_{231}	$\bar{y}_{23.}$	y_{241}	$\bar{y}_{24.}$
		y_{212}		y_{222}		y_{232}		y_{242}	
	3	y_{311}	$\bar{y}_{31.}$	y_{321}	$\bar{y}_{32.}$	y_{331}	$\bar{y}_{33.}$	y_{341}	$\bar{y}_{34.}$
		y_{312}		y_{322}		y_{332}		y_{342}	
	4	y_{411}	$\bar{y}_{41.}$	y_{421}	$\bar{y}_{42.}$	y_{431}	$\bar{y}_{43.}$	y_{441}	$\bar{y}_{44.}$
		y_{412}		y_{422}		y_{432}		y_{442}	
	5	y_{511}	$\bar{y}_{51.}$	y_{511}	$\bar{y}_{52}$	y_{531}	$\bar{y}_{53.}$	y_{541}	$\bar{y}_{54.}$
		y_{512}		y_{522}		y_{532}		y_{542}	

Factor A (weight): has five levels that are:

I. empty car plus driver's weight
II. empty car plus driver's and another person
III. empty car plus driver's and two other persons
IV. empty car plus driver's and three other persons
V. empty car plus driver's and four other persons

Factor B (acceleration): has 4 levels that are:
The values of this factor have been selected according to the acceleration that is in range of factory information:

I. $2.1^{(m/s^2)}$
II. $3.1^{(m/s^2)}$
III. $4.1^{(m/s^2)}$
IV. $5.1^{(m/s^2)}$

3.2.1 Formulate effects model and test hypotheses

The observation in a factorial experiment can be described by a model. In this experiment all factors are fixed, therefore it is easy to test hypotheses and formulate effects model. Also the probability of type one error, α , often called the significance level had chosen as 0.05 in this experiment.

For a fixed effect model, test statistics for each main effect and interaction may be constructed by dividing corresponding mean square error. All of these F test will be upper tail, one tail test. The number of degree of freedom for any main effect is the number of levels of the factor minus one, and the number of degree of freedom for an interaction is the product of the number of degrees of freedom associated with the individual components of the interaction. Effects model and testing hypothesis in the following pages.

In this design a is the number of weights that has been used in this research, and b is the number of accelerations that has been assumed for the car. The total of each treatment can be shown in table 3.

Table 3 Totals of each treatment (values in dark cells are average of each interaction treatment)

weight (kg)	acceleration (m/s²) 2.1		3.1		4.1		5.1		$y_{i.}$
827	4.037	**4.196**	2.940	**3.381**	2.312	**2.421**	1.905	**2.065**	24.125
	4.356		3.821		2.529		2.225		
902	3.752	**4.222**	2.723	**3.223**	2.137	**2.153**	1.758	**2.221**	23.634
	4.691		3.722		2.168		2.683		
977	3.504	**3.856**	2.535	**2.694**	1.986	**2.480**	1.632	**1.879**	21.814
	4.207		2.852		2.973		2.125		
1052	3.287	**3.371**	2.371	**2.408**	1.855	**1.922**	1.523	**1.600**	18.598
	3.454		2.444		1.988		1.676		
1127	3.096	**3.311**	2.228	**2.289**	1.740	**2.099**	1.427	**1.713**	18.822
	3.525		2.350		2.458		1.998		
$y_{.j.}$	37.909		27.986		22.146		18.952		$y_{...}$=106.993

4 Statistical Analysis

4.1 Analysis of the data

Statistical methods should be used to analyze the data so that results and conclusion are objective rather than judgmental in nature. If the experiment has been designed correctly and if it has been performed according to the design, the statistical methods required are not elaborate. There are many excellent software packages designed to assist in data analysis.

In this paper, **Mathematica** and **MS- Excel** has been used to analysis the data. Tables and formulas that construct in the following pages make everything clear to realize steps of analyzing.

4.2 Analysis of variance (ANOVA)

Mathematica and MS-Excel for constructing analysis of variance have been used.

The analysis of variance is summarized in table 4 that constructed with Mathematica software package. The primary assumption in this work is $\alpha = 0.01$. From the table 4 and the assumption of the α value we can understand that none of the two factors, weight and acceleration, lonely is significantly affecting the fuel consumption of the car. But the interaction of these two factors is significantly affecting the rate of fuel consumption.

Table 4 analysis of variance table

Treatment	SS	DOF	MS	F	P-Value
weight	3.385	4	0.84625	5.038702	0.005545
acceleration	20.805	3	6.935	41.29205	$9.031*10^{-8}$
Interaction of weight and acceleration	0.95	12	0.079167	0.47137	0.903133
error	3.359	20	0.16795		
total	28.499	39			

However, to compare the results of each treatment, Tukey method was used to show that each treatment is significantly different from the others or not. To know the answer of this question each pair of the treatments together with the Tukey method is compare as bellow.

The rate of fuel consumption in this project has been shown as unit value of (km/lit). This means that the best fuel consumption is that value which gives maximum travel distance passed by the car under the same fuel consumed.

PostTests → {Weight
→ Tukey {{827,1052},{902,1052},{827,1127}}

According to the above statement that is from Mathematica software, we can understand that there is no significant difference between weight 827, 902, and 977 kg. This shows that we can run the car with 977 kg weight without sensible change in fuel consumption in compare with the time that we run it with 827 kg weight. Logically we know that the optimum usage of the car is in the time that we use it with more weight.

PostTests → {Acceleration
→ Tukey {{2.1,3.1},{2.1,4.1},{3.1,4.1},{2.1,5.1},{3.1,5.1}}

According to the above statement that is from Mathematica software, we can understand that there is significant difference between acceleration 2.1 and other accelerations that have been experienced in this project and also we see that the more acceleration, the more fuel consumption is caused. Then the economical acceleration of the car is 2.1 meter per square second.

4.2.1 Analysis of residual

After performing analysis of variance, the second step is to check of the normality assumption that could be made by plotting residual histogram. useful procedure is to construct a normal probability plot of the residuals. By using following formulas we can find the residuals, and data of residuals summarized in table 5.

$$e_{ijk} = y_{ijk} - \hat{y}_{ijk} \quad \& \quad \hat{y}_{ijk} = \bar{y}_{ij.} \Rightarrow e_{ijk} = y_{ijk} - \bar{y}_{ij.}$$

Also in Fig. 2, normality assumption had checked and as it is shown in this figure there is no point outlier. The reason is that, the data have been chosen randomly & it has variety of scale (quantity).

Table 5 residual versus predicted values

y_{ijk}	e_{ijk}	y_{ijk}	e_{ijk}	y_{ijk}	e_{ijk}	y_{ijk}	e_{ijk}
4.04	-0.16	2.94	-0.44	2.31	-0.11	1.91	-0.16
4.36	0.16	3.82	0.44	2.53	0.11	2.23	0.16
3.75	-0.47	2.72	-0.50	2.14	-0.02	1.76	-0.46
4.69	0.47	3.72	0.50	2.17	0.02	2.68	0.46
3.50	-0.35	2.54	-0.16	1.99	-0.49	1.63	-0.25
4.21	0.35	2.85	0.16	2.97	0.49	2.13	0.25
3.29	-0.08	2.37	-0.04	1.86	-0.07	1.52	-0.08
3.45	0.08	2.44	0.04	1.99	0.07	1.68	0.08
3.10	-0.21	2.23	-0.06	1.74	-0.36	1.43	-0.29
3.53	0.21	2.35	0.06	2.46	0.36	2.00	0.29

Furthermore in this experiment there is a fixed factor design model so it is possible to use the data without eliminating any data or repeating any experiment.

4.2.2 Test for equality of variance

Another assumption that must be satisfied for this experiment is equality of variance. If the model is correct and if the assumptions are satisfied, the residual should be structure less. In this experiment check equality of variance by plotting residual versus the predicted values $\hat{y}_{ijk} = \bar{y}_{ij.}$.

As shown in the Fig. 3, there is no obvious pattern so we can conclude that equality of variance satisfied and assumption of homogeneity of variance is not violated.

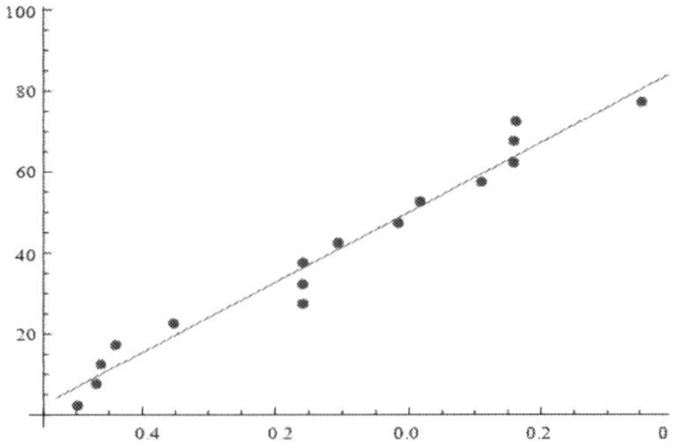

Fig. 2 Normal probability plot of residuals

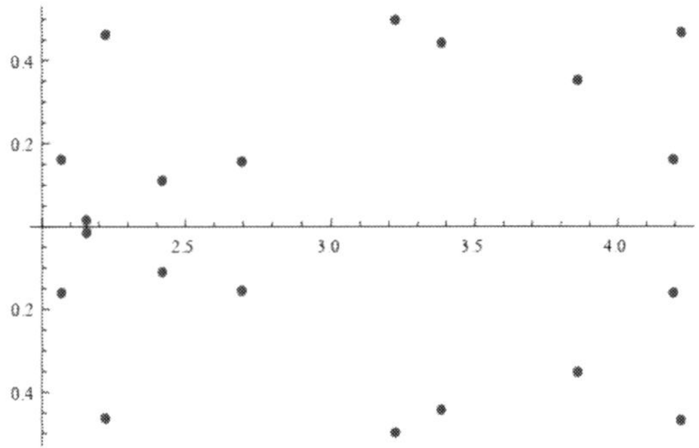

Fig. 3 residual versus predicted values

5 Conclusions

The effect of acceleration & weight of a car on its fuel-consumption had been studied using Design and Analysis of Experiments. By using ANOVA it was shown that the fuel consumption increases with the increase of the car weight. This means, as the mass of the car increases, the travel distance of the car per unit volume of the fuel will decreases and accordingly, the cost per unit travel of the car will increase. As a result of this relation, it seems sensible to explore every avenue towards saving energy in cars making.

Mathematica software was used to obtain this decision in this experiment. The conclusions that have been drawn from the analysis are summarized as follow:

1. The maximum persons that can be in the car without affecting the fuel consumption significantly, is three.
2. The maximum acceleration that is economically is 2.1 meter per square second.

Acknowledgement

The author would like to thank Assoc. Prof. Dr. Adham Makeia, Department of Industrial Engineering, Eastern Mediterranean University, turkey for his help and support with various issues during doing the work.

References

1. Royce, king C.S.,"A car for the nineties", BL's energy conservation vehicle, transport engineering, part D, 1984, Vol. 198, No.11, PP. 21-25.
2. Walid. M. Al-Momani, Omar. O. Badran, Experimental Investigation of Factors Affecting Vechicle Fuel Consumption. International Journal of Mechanical and Materials Engineering (IJMME), 2007,Vol. 2 ,No. 2, 180—188
3. Charles H Heide and Pirooz Mohazzabi, Fuel economy of a vehicle as a function of airspeed: the concept of parallel corridors. International Journal of Energy and Environmental Engineering vol, 2013,Vol 4, No.2
4. Uldis Putnieks, Automobile Fuel Consumption Changes As Effect of Load. Engineering for Rural Development, Jelgava,2013, Vol.5
5. Jan Rouwendal, An Economic Analysis of Fuel Use Per Kilometer By Private Cars. Journal of Transport Economics and Policy.(1996).
6. A shly. M.F. & Jones R.H. Engineering materials. Materials & energy in car design, 1980,vol. 34, PP.239-248.
7. D. G. Montgomery, Design and analysis of experiments, Fifth Edition, John Wihley and Sons Inc., 2001.

AUTHOR INDEX

Ali Hussein M. 98
Anicas John Mark S. 71
Aquino Ruth R. 84
Arcamo Niel Karl G. 84
Baothong Thanakhan 60
Basilia Blessie A. 84
Cai Biao-Hua 93
Cayme Jan-Michael C. 71, 75
Cheng Jun 50
Fang Chao 93
Feng Huihui 9
Fu Qiang 79
Gara John Patrick N. 84
Han Bing 79
Han Ningda 50
Hu Luo 64
Huang Jie 14
Huang Shyh-Chour 41
Huang Wuyun 50
Huang Yu-Che 5
Hung Le Xuan 36
Iqbal Asif 98
Jiang Zhengquan 14
Jun Liu 1
Lai Lei-Jie 22
Laplana Nina Muriel A. 75
Li Bin 93
Li Zhang 1
Liu Haitao 26
Liu Yijian 50
Lo Jih-Sheng 41
Ma Shi-Hua 93
Mangay John Carlo A. 71, 75
Meiyun Yu 1
Miranda Eric T. 71, 75
Nguyen Huu Loc 30
Osvaldová Linda Makovická 89
Peng Liping 9
Pi Vu Ngoc 36
Pitayachaval Paphakorn 46, 55, 60
Recto Angel S. 71, 75
Sanklong Nattawut 55
Sato Muhammatsoifu 46
Sun Shuwen 14
Tang Jianjun 26
Tham Hoang Thi 36
Thanh Ho Ky 36
Thao Tran Thi Phuong 36
Thi Hong Cam Nguyen 36
Thongrak Anantapoom 55
Tolentino Marvin S. 84
Tuan Nguyen Khac 36
Van Thuy Tran 30
Yafeng Ren 1
Yang Jianwu 14
Yang Jiquan 50
Yanyan Luo 1
Yeh Tien-Chaun 5
Yin Xiao-Qia 22